SAMPSON COMMUNITY COLLEGE LIBRARY

Theoretical and Practical Ammonia Refrigeration

Iltyd I. Redwood

BIBLIOBAZAAR

Copyright © BiblioBazaar, LLC

BiblioBazaar Reproduction Series: Our goal at BiblioBazaar is to help readers, educators and researchers by bringing back in print hard-to-find original publications at a reasonable price and, at the same time, preserve the legacy of literary history. The following book represents an authentic reproduction of the text as printed by the original publisher and may contain prior copyright references. While we have attempted to accurately maintain the integrity of the original work(s), from time to time there are problems with the original book scan that may result in minor errors in the reproduction, including imperfections such as missing and blurred pages, poor pictures, markings and other reproduction issues beyond our control. Because this work is culturally important, we have made it available as a part of our commitment to protecting, preserving and promoting the world's literature.

All of our books are in the "public domain" and many are derived from Open Source projects dedicated to digitizing historic literature. We believe that when we undertake the difficult task of re-creating them as attractive, readable and affordable books, we further the mutual goal of sharing these works with a larger audience. A portion of Bibliobazaar profits go back to Open Source projects in the form of a donation to the groups that do this important work around the world. If you would like to make a donation to these worthy Open Source projects, or would just like to get more information about these important initiatives, please visit www.bibliobazaar.com/opensource.

Theoretical and Practical Ammonia Refrigeration

Iltyd I. Redwood

BIBLIOBAZAAR

Copyright © BiblioBazaar, LLC

BiblioBazaar Reproduction Series: Our goal at BiblioBazaar is to help readers, educators and researchers by bringing back in print hard-to-find original publications at a reasonable price and, at the same time, preserve the legacy of literary history. The following book represents an authentic reproduction of the text as printed by the original publisher and may contain prior copyright references. While we have attempted to accurately maintain the integrity of the original work(s), from time to time there are problems with the original book scan that may result in minor errors in the reproduction, including imperfections such as missing and blurred pages, poor pictures, markings and other reproduction issues beyond our control. Because this work is culturally important, we have made it available as a part of our commitment to protecting, preserving and promoting the world's literature.

All of our books are in the "public domain" and many are derived from Open Source projects dedicated to digitizing historic literature. We believe that when we undertake the difficult task of re-creating them as attractive, readable and affordable books, we further the mutual goal of sharing these works with a larger audience. A portion of Bibliobazaar profits go back to Open Source projects in the form of a donation to the groups that do this important work around the world. If you would like to make a donation to these worthy Open Source projects, or would just like to get more information about these important initiatives, please visit www.bibliobazaar.com/opensource.

THEORETICAL AND PRACTICAL

Ammonia Refrigeration.

A WORK OF REFERENCE FOR ENGINEERS
And others Employed in the Management of Ice and Refrigeration Machinery.

BY

ILTYD I. REDWOOD,

ASSOC. M. AM. SOC. OF M. E.: M. SOC CHEMICAL INDUSTRY, ENGLAND.

WITH 25 PAGES OF TABLES.

NEW YORK:
SPON & CHAMBERLAIN, 12 CORTLANDT STREET.
LONDON:
E. & F. N. SPON, 125 STRAND.
1895.

PREFACE.

THERE are many engineers and others interested in refrigerating machinery who have felt the want of a book of reference that will enable them to determine, with sufficient accuracy for all practical purposes, what work their machines are doing without resorting to laborious calculations; therefore a number of tables have been prepared to meet this want, and a short treatise on the Theory and Practice of Refrigeration incorporated therewith.

The tables, which have been calculated as accurately as possible, and have been checked by a gentleman of considerable "expert" experience, cover a sufficiently wide range of temperatures and pressures to meet all ordinary, and a good many extraordinary, requirements.

ILTYD I. REDWOOD.

BROOKLYN, *February*, 1895.

CONTENTS.

	PAGE
INTRODUCTORY REMARKS	1

CHAPTER I.

BRITISH THERMAL UNIT	3
MECHANICAL EQUIVALENT OF A UNIT OF HEAT	4
SPECIFIC HEAT	4
EFFECT OF TEMPERATURE AND PRESSURE ON SPECIFIC HEAT	6
EFFECT OF PRESSURE ON SPECIFIC HEAT OF AMMONIA GAS	7
SPECIFIC HEAT OF AIR WITH CONSTANT PRESSURE	7
SPECIFIC HEAT OF AIR WITH CONSTANT VOLUME	9
LATENT HEAT	10
LATENT HEAT OF LIQUEFACTION	10
LATENT HEAT OF VAPORIZATION	11
LATENT HEAT OF WATER	12
ABSOLUTE PRESSURE	13
ABSOLUTE TEMPERATURE	13
ABSOLUTE ZERO	16
EFFECT OF PRESSURES ON VOLUME OF GASES	16

Contents.

CHAPTER II.

	PAGE
THEORY OF REFRIGERATION	18
FREEZING BY COMPRESSED AIR	19
FREEZING BY AMMONIA	21
CHARACTERISTICS OF AMMONIA	22
EXPLOSIVENESS	23
TENDENCY OF THE GAS TO RISE	24
SOLUBILITY IN WATER	24
ACTION ON COPPER	25
26° BEAUMÉ AMMONIA	25
ANHYDROUS AMMONIA	25

CHAPTER III.

GENERAL ARRANGEMENT	26
DESCRIPTION OF THE PLANT	27
CONSTRUCTION DETAILS—THE COMPRESSOR	30
STUFFING-BOXES	32
SPECIAL LUBRICATION	34
OIL FOR LUBRICATION	35
CLEARANCE SPACE, ETC.	35
SUCTION AND DISCHARGE VALVES	36
EFFECT OF EXCESSIVE VALVE-LIFT	37
REGULATION OF VALVE-LIFT	37

CHAPTER IV.

THE SEPARATOR	38
THE CONDENSER	42
CONDENSER-WORM	42
RECEIVER	43

Contents.

iii.

	PAGE
REFRIGERATOR OR BRINE TANK	44
SIZE OF PIPE AND AREA OF COOLING SURFACE	45
EXPANSION VALVES	46
WORKING DETAILS.—CHARGING THE PLANT WITH AMMONIA	47

CHAPTER V.

AMMONIA TO BE GRADUALLY CHARGED	49
JACKET-WATER FOR COMPRESSOR	52
JACKET-WATER FOR SEPARATOR	53
CONDENSING WATER	53
LESSENING THE COST FOR CONDENSING WATER	54
QUANTITY OF CONDENSING WATER NECESSARY	56
LOSS DUE TO HEATING OF CONDENSED AMMONIA,	56
LOSS DUE	58
SUPERHEATING AMMONIA GAS	58

CHAPTER VI.

EXCESS CONDENSING PRESSURE	59
CAUSE OF VARIATION IN EXCESS PRESSURES	60
OTHER CONDITIONS THAT AFFECT EXCESS PRESSURE,	62
USE OF CONDENSING PRESSURE IN DETERMINING LOSS OF AMMONIA BY LEAKAGE	63
COOLING DIRECTLY BY AMMONIA	65
BRINE	66
FREEZING POINT OF BRINE	68
EFFECT OF COMPOSITION ON FREEZING POINT	68
EFFECT OF STRENGTH ON FREEZING POINT	69
SUITABLENESS OF THE BRINE	70
MAKING BRINE	71

CHAPTER VII.

	PAGE
SPECIFIC HEAT OF BRINE	73
REGULATION OF BRINE TEMPERATURE	73
INDIRECT EFFECT OF CONDENSING WATER ON BRINE TEMPERATURE	77

CHAPTER VIII.

DIRECTIONS FOR DETERMINING REFRIGERATING EFFICIENCY	78
EQUIVALENT OF A TON OF ICE	79
COMPRESSOR MEASUREMENT OF AMMONIA CIRCULATED	79
LOSS IN WELL-JACKETED COMPRESSORS	80
LOSS IN DOUBLE-ACTING COMPRESSORS	80
DISTRIBUTION OF MERCURY WELLS	81
EXAMINATION OF WORKING PARTS	86
NUMBER OF READINGS TO BE TAKEN	86

CHAPTER IX.

DURATION OF TEST	87
INDICATOR DIAGRAMS	87
AMMONIA FIGURES.—EFFECTUAL DISPLACEMENT	97
VOLUME OF GAS	97
AMMONIA CIRCULATED PER TWENTY-FOUR HOURS	98
REFRIGERATING EFFICIENCY	98
BRINE FIGURES.—GALLONS CIRCULATED	99
POUNDS CIRCULATED	100
DEGREES COOLED	100
TOTAL DEGREES EXTRACTED	100

CHAPTER X.

	PAGE
Loss Due to Heating of Liquid Ammonia	102
Loss Due to Heating of Ammonia Gas	103

CHAPTER XI.

Calculation of the Maximum Capacity of a Machine	106
Preparation of Anhydrous Ammonia	107
Construction of Apparatus	108
Condenser-Worm	109
Why Still is Worked under Pressure	110
Best Test for Ammonia	111
Water from Separators	101
Lime for Dehydrator	111
Yield of Anhydrous from 26° Ammonia	112

Index 139

List of Illustrations.

Fig.		Page
1.	Specific Heat with Constant Pressure Determination	8
2.	Absolute Zero Determination	14
3.	Ammonia Plant	28
4.	" "	29
5.	Discharge Valve	36
6.	Suction "	36
7.	Separator	40
8.	Expansion Valve	46, 47
9.	Mercury Well	82
10.	" "	84
11.	Indicator Diagram	88
12.	" "	89
13.	" "	90
14.	" "	91
15.	Anhydrous Ammonia Distilling Apparatus	115

Tables.

Table		Page
I.	Volume of Ammonia Gas at High Temperatures,	51
II.	Yield, etc., of Anhydrous Ammonia from Ammonia Solutions	113
III.	Boiling Point, Latent Heat, etc., of Anhydrous Ammonia	116, 117
IV.	Temperature to which Ammonia Gas is raised by Compression	118 to 122
V.	Volume of One Pound of Ammonia Gas at Various Pressures and Temperatures,	122 to 130
VI.	Volume of One Pound of Ammonia Gas at Various Pressures and Temperatures,	131 to 138

AMMONIA REFRIGERATION.

Introductory Remarks.

The ammonia "compression" types of freezing machines are now coming so generally into use in large factories and manufacturing establishments where natural ice was formerly employed, that they are of necessity placed directly or indirectly under the supervision of men who, owing to the comparative newness of the subject of ammonia refrigeration in relation to the manufactures, can not be expected to be thoroughly conversant with their theoretical and practical working.

In a great many instances engineers who have charge of these machines only run them by rule-of-thumb methods, and knowing nothing about the why and the wherefore are, in the event of the conditions being changed, unable to reason out what will result from the changed conditions, and what other changes ought to be made to counterbalance them.

It is therefore with a view to giving those connected with the running of ammonia refrigerating plants a more intelligent idea of what they are doing—thereby tending to make their work interesting instead of laborious—that this Book has been written.

CHAPTER 1.

BEFORE dealing with ammonia refrigeration it is necessary that the different heat terms, etc., that are used in regard to this subject should be thoroughly understood, and they will therefore be explained forthwith.

The terms with which we have principally to deal are :

(1) British Thermal Unit.
(2) Mechanical Equivalent of a Unit of Heat.
(3) Specific Heat.
(4) Latent Heat.
(5) Absolute Pressure.
(6) Absolute Temperature.

British Thermal Unit.

A British thermal unit is the standard unit of heat in this country, and represents the amount of heat necessary to raise the temperature of one pound weight of water one degree Fahrenheit—the temperature of the water being 32°; on the other hand, it is the amount of heat given up by one pound of water in cooling one degree Fahrenheit (*i. e.*, from 33° down to 32°).

Mechanical Equivalent of a Unit of Heat.

Joule found, by means of a suitably constructed agitator placed in water and actuated by a falling weight, that the amount of friction caused by a weight of 1 lb. falling a distance of 772 feet, or a weight of 772 lbs. falling a distance of 1 foot, was sufficient to heat 1 lb. of water 1° Fahr. Therefore, the production of one British thermal unit of heat is equivalent to raising a weight of 1 lb. 772 feet, or 772 lbs. 1 foot, and consequently the mechanical equivalent of a unit of heat is 772 foot-pounds.

Specific Heat.

Specific heat is the number of British thermal units required to raise the temperature of one pound weight of any particular substance 1° Fahr., or it may be expressed as the capacity of different substances for heat.

Scientists have proved that a pound of water has a greater capacity for heat than a pound of any other known substance, and therefore water is taken as the standard of comparison, and its specific heat at 32° Fahr. is unity.

Turpentine has a specific heat of 0.472 and the specific heat of mercury is 0.033; from these figures it is understood that to raise the temperature of 1 lb. of turpentine 1° Fahr. 0.472 B. T. U.* will be required, while the same weight of mercury will require only 0.033 B. T. U. to raise its temperature one degree.

If 2 lbs. of water at 32° Fahr. are heated to 42° Fahr., or through 10°, they will absorb (2 lbs. × 10° × 1.000 Sp. Ht. =) 20 B. T. U's, but if 2 lbs. of turpentine are heated through the same number of degrees they

---

* British Thermal Units.

will absorb only (2 lbs. × 10° × 0.472 Sp. Ht. =) 9.44 B. T. U's.

Effect of Temperature and Pressure on Specific Heat.

The specific heat of substances varies with varying conditions of temperature and pressure, and invariably increases with increase of temperature or pressure. The variation in the specific heat of water at different temperatures is so small that it may be passed unnoticed, but in the cases of certain oils and gases it is considerable: for instance, a mineral oil that has a specific heat of 0.4503 at 85° Fahr. will have a specific heat of 0.4843 at 120° Fahr. Another point in regard to the specific heat of mineral oils is the fact that as the weight (specific gravity) of the oil "increases" the specific heat "decreases." Also, in the case of paraffin waxes, the higher the melting point the lower the specific heat.

Effect of Pressure on Specific Heat of Ammonia Gas.

The effect of pressure on the specific heat of ammonia gas is very marked, for whereas the specific heat is only 0.508 when the gas is under a pressure of 28 lbs. or less on the square inch, it is raised to 0.532 when the pressure reaches 80 lbs. or upwards.

The specific heat of a gas when expansion is allowed and when mechanical work is performed is greater than the specific heat of a gas that is not allowed to expand; in other words, specific heat of a gas with constant pressure is greater than the specific heat of a gas with constant volume. In order to understand this more clearly, the following explanation must be given:

Specific Heat of Air with Constant Pressure.

Let Figure 1 represent a cylinder with a cross sectional area of 144 square inches (one square foot) tightly closed at both ends and fitted with a piston, B, that will move without friction, and let the piston weigh 2,116.2 lbs. Now, if a perfect vacuum is maintained in the space A, and if C contains 1 lb. of air (= 12.387 cubic feet) at a temperature of 32° Fahr., the air will be under a pressure of 14.696 lbs. per square inch, and will maintain the piston at a height of 12.387 feet. If this air is now heated to 33° Fahr.—thus raising its temperature 1° Fahr.—its volume will be increased, but the pressure will be exactly the same as before, because the piston has risen to make room for the increased volume of the air. According to Regnault's determinations, the amount of heat that would be necessary to raise the temperature of the air 1° Fahr. under the above conditions, would be 0.2379 B. T. U. Therefore the specific heat of air with constant pressure is 0.2379.

Fig. 1

Specific Heat of Air with Constant Volume.

In the experiment just cited, not only was the temperature of the air raised 1° Fahr., but, owing to its expansion, a certain amount of mechanical work was performed when the piston was raised. Now, by heating the air 1° Fahr., its volume was increased (see page 16) to $\left(12.387 \times \dfrac{458.4 + 33°}{458.4 + 32°} =\right)$ 12.41226 cubic feet, therefore the piston was raised from 12.387 feet up to 12.41226 feet, or through 0.02526 of a foot. As already mentioned, the piston weighed 2,116.2 lbs., therefore the amount of work done by the expansion of the air was 2,116.2 lbs. × 0.02546, height raised = 53.4552 foot-pounds. As it is known that the mechanical equivalent of a unit of heat is 772 foot-pounds, it is seen that the amount of heat that was required to perform the mechanical work of raising the piston was 53.4552 ÷ 772 = 0.06924 B. T. U. Therefore, if the air had been heated from 32° up to 33° Fahr. without being allowed to expand and perform mechanical work, the amount of heat that would have been necessary would have been (0.2379 − 0.06924 =) 0.16866 B. T. U.; hence the specific heat of air with constant volume is 0.16866.

LATENT HEAT.

Latent heat is heat that is hidden or is absorbed (without making itself apparent to the thermometer) when a solid passes to the liquid state, or a liquid to the gaseous state.

There are, therefore, two kinds of latent heat, one being the latent heat of liquefaction and the other the latent heat of vaporization.

LATENT HEAT OF LIQUEFACTION.

If 1 lb. of ice at 32° Fahr. and 1 lb. of water at 33° Fahr. are placed in separate vessels of exactly the same size and shape, and these vessels are put in a place that is perfectly free from draughts and where the temperature is stationary at, say, 50° Fahr., it will be found that the ice will take about 21 times as long to melt and heat up to, say, 40° Fahr. as the water will take to heat up to the same temperature. Now it is quite plain that if both vessels are exposed to exactly the same temperature, their contents must each be absorbing heat at the same rate, and as the temperature of the water in rising from 33° to 40°, or through seven degrees, only required 1-21st of the time that the ice took, the ice must have absorbed (7 × 21) = 147° Fahr., but only 8° (32° to 40°) of this had been registered by the thermometer, and therefore 139° Fhr. had become latent or hidden. Of course this is but a crude method of determining latent heat, and accurate determinations have fixed 142.4 as the latent heat of ice.

LATENT HEAT OF VAPORIZATION.

If water is heated in an open vessel it will be found that the temperature can not be raised above 212° Fahr. No matter how long the heat may be applied the temperature will remain stationary, although the water is constantly receiving additional heat. The heat thus hidden in the water is called the latent heat of vaporization, and if 1 lb. of steam at 212° Fahr. were passed through a condenser and converted into 1 lb. of water at 212° Fahr. it would be found that, although the condensation of the steam to water had not affected the temperature sufficiently to be noticeable by the thermometer, the condenser would have absorbed 966 B. T. U's, or sufficient heat to have raised the temperature of over $6\frac{1}{4}$ lbs. of water from 60° Fahr. up to 212° Fahr.

The latent heat of vaporization of water is therefore 966.

LATENT HEAT OF WATER.

It is thus seen that to convert 1 lb. of ice at 32° Fahr. into 1 lb. of steam at 212° Fahr. requires:

Ice at 32° to water at 32° (latent) . .	142.4
Water at 32° to water at 212° . . .	180.0
Water at 212° to steam at 212° (latent)	966.0
	1,288.4 B. T. U's;

or the amount of heat that would reduce about 2½ lbs. of cast-iron or about 9 lbs. of silver to the molten state.

In making a great many calculations in regard to heat it is necessary to make use of absolute pressures and temperatures.

Absolute Pressure.

Absolute pressure is pounds per square inch above a vacuum, and, as steam gauges are adjusted so that the O, or zero mark, represents the atmospheric pressure, it is necessary to add 14.7 lbs. to the guage pressure, in order to convert it into absolute pressure.

Absolute Temperature.

In regard to absolute temperature experiments have proved that all pure, dry gases expand very nearly to the same extent for equal increments of heat, and it therefore matters little what gas is taken for the purpose of explaining the principle on which the basis for absolute temperatures has been determined.

Let Fig. 2 be a cylinder closed at both ends, and having a cross sectional area of 144 square inches (1 square foot), a depth of about 18 inches, and a piston, B, capable

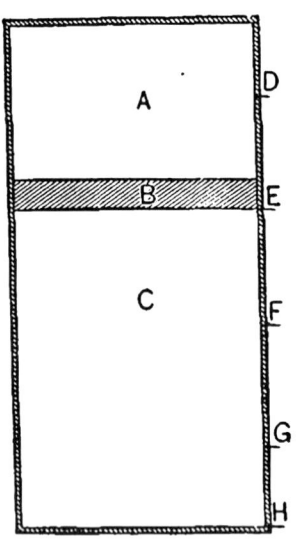

Fig. 2

of moving without friction. It must now be supposed that the space C contains 1 cubic foot of air at a temperature of 32° Fahr., and that the piston, B, is weighted so as to exert a pressure of 14.7 lbs. on the square inch, while a perfect vacuum is maintained in A. Regnault's experiments have proved that if the contents of C are now heated to 212° Fahr., or through 180° Fahr. (*i. e.*, 212° −32°), the piston and its load will be raised 0.367 foot, or to D, and the cubic foot of air will be increased in volume to 1.367 cubic feet.

If we start again with the temperature at 32° Fahr. and the piston at E, and extract instead of add 180° Fahr. of heat (*i.e.*, cool down the contents of C to − 148° Fahr.), the piston will descend the same distance that it rose when the air was heated, namely, 0.367 foot, or to F. The extraction of another 180° Fahr. by cooling down the contents of C to −328° Fahr., would cause the piston to again descend another 0.367 foot, or to G, and to cause the piston to descend to H (and thus contract the air in C to, theoretically speaking, nothing), would necessitate the air being cooled down $\frac{180}{367}$ = 490.4° Fahr. below 32° Fahr. or to 458.4° Fahr. below zero.

Absolute Zero.

Absolute zero is − 458.4° Fahr., and an absolute temperature is the absolute zero temperature, plus the ordinary thermometer reading. The absolute temperature of a gas at 32° Fahr. is 490.4 (458.4 + 32°), and if the temperature were 0° Fahr. the absolute temperature would be 458.4, while if the temperature were − 32° the absolute temperature would be 426.4 (= 458.4 − 32°).

With the aid of this knowledge it is now easy to understand how the volume of gases at different temperatures is computed by the formula $v = V \times \dfrac{458.4 + t}{458.4 + T}$, in which

V = Volume of the gas at the original temperature, T.

v = volume of the gas at the new temperature, t.

Effect of Pressures on Volume of Gases.

The volume of gases is also altered by pressure, and, according to Marriotte, the volume of any gas varies inversely as the pressure—the temperature remaining constant. Thus: one cubic foot of air at 10 lbs. absolute pressure on the square inch, if subjected to an absolute pressure of 100 lbs., will be reduced in volume to (1 cubic foot × 10 lbs. ÷ 100 lbs. =) 0.1 cubic foot, provided the work of compressing is done without generating heat. But it is known that when work is done, heat is necessarily generated, and if the cubic foot of air at 10 lbs. absolute pressure is compressed to 1-10th its volume by being subjected to an absolute pressure of 100 lbs., its temperature will be raised to about 810° Fahr. Therefore, in calculating the volume of a gas that has been subjected to pressure, it is necessary to take into consideration the changes in volume caused by both temperature and pressure together, and the general formula becomes:

$$V \times \frac{P}{p} \times \frac{458.4 + t}{458.4 + T} = v,$$

in which V, P and T, and v, p and t, are the respective volumes, pressures and temperatures of the gas before and after compression. Thus, if

1 cubic foot of air.................................. = V
at 20 lbs. Absolute Pressure....................... = P
and 60° Fahr. temperature......................... = T
   is heated to
600° Fahr. temperature............................ = t
   by being subjected to
200 lbs. Absolute Pressure........................ = p
   it will be reduced in volume to:

$$1 \text{ cubic foot} \times \overset{Pres.}{\frac{20}{200}} \times \overset{Temp.}{\frac{458.4 + 600}{458.4 + 60}} = 0.2 \text{ cubic ft} \ldots\, v$$

CHAPTER II.

THEORY OF REFRIGERATION.

A CAREFUL study of the foregoing pages ought to have made the two following facts quite plain:

1. In order to effect the expansion of a

gas it is necessary that the gas should absorb heat.

2. The act of compressing a gas generates heat.

Freezing by Compressed Air.

If a compressed gas is re-expanded it practically absorbs the same amount of heat that was generated by compression, and the re-expanded gas will therefore be cooled down to its original (*i. e.*, before compression) temperature. The gas in this case will simply absorb the heat necessary for its re-expansion from itself; but if, on the other hand, the compressed gas is cooled down before it is allowed to re-expand, it is very evident that it will not contain sufficient heat in itself to effect its own expansion, and therefore it will have to extract the necessary heat from its surroundings, and by so doing it will produce the sensation of cold, although, strictly speaking, cold can not be produced, as it is a negative condition.

Theoretical and Practical

The following example will make the foregoing explanation plainer:

1 lb. of air at......................14.7 lbs. Abs. Pres.
    and...........................60° Fahr.
if compressed to....................110 lbs. Abs. Pres.
will have its temperature raised to..475° Fahr.
This compressed air is now cooled to.65° Fahr.
or through......(475° − 65°)........410° Fahr.
As the specific heat of air is 0.238,
    the number of thermal units that
    have been extracted from the compressed air are...(410 × 0.238)....97.58.

If this cool compressed air is now re-expanded to its original absolute pressure of 14.7 lbs., it will have to absorb 97.58 B. T. U's. As the extraction of 170 thermal units from 1 lb. of water whose temperature is 60° Fahr. will convert the pound of water into a pound of ice, it is evident that if the 1 lb. of above compressed air at a temperature of 65° Fahr. is expanded in a suitable apparatus surrounded by (97.58 ÷ 170 =) 0.574 lb. of water at 60° Fahr. temperature, the water will be converted into 0.574 lb. of ice of 32° Fahr. temperature.

The above figures are only approximately

correct, and are simply given as an illustration of the theory of freezing by compressing and re-expanding a gas (such as air) that is not liquefied by compression.

Freezing by Ammonia.

In considering the theory of refrigeration by means of the liquefiable gas ammonia it will be seen that the great advantage of ammonia over air lies almost entirely in the latent heat of vaporization.

Suppose 1 lb. of ammonia gas at 20 lbs. absolute pressure and 32° Fahr. is compressed to 110 lbs. absolute pressure, its temperature will thereby be raised to 268.6° Fahr. If the compressed gas is cooled to 65° Fahr. its temperature will be lowered 203.6°, and this number of degrees multiplied by the specific heat of ammonia gas (which in this case is 0.532) shows that 108.31 thermal units have been extracted from the gas. But if instead of cooling the compressed gas to only 65° Fahr. it is cooled to 60° Fahr., it will be converted into a liquid, and as the latent heat of vaporization of ammonia at 110 lbs. absolute pressure is 517.23, the following will now be the number of thermal units extracted. Temperature of compressed gas was 268.6° Fahr., and if cooled to 60° Fahr. its temperature will be lowered 208.6°.

Degrees cooled × specific heat........ = 110.97 T. U's.
Latent heat of vaporization............ = 517.23 "
_____
Therefore total thermal units extracted = 628.20

These figures show how the advantage derived by the use of ammonia in the place of air lies in the comparative ease with which ammonia gas can be liquefied, thereby allowing of use being made of its latent heat of vaporization.

CHARACTERISTICS OF AMMONIA.

Ammonia is a colorless, irrespirable gas, with the odor of hartshorn. It is feebly combustible if mixed with a large proportion of air, and burns with a greenish-yellow flame; if mixed with about twice its volume of air it explodes with some violence. It is only a little more than half the weight of air, is exceedingly soluble in water, and has a very strong action on copper and its alloys. The characteristics of ammonia render it necessary that the following precautions should be observed in regard to the handling of it and in constructing an ammonia refrigerating plant.

EXPLOSIVENESS.

Owing to the explosiveness of the gas it is important that any part of an apparatus should be thoroughly aired before a naked light is brought near it. This precaution is sometimes ridiculed by those who, through good luck rather than good management, have never exploded any large volume of the gas; but the author has personal knowledge of a case where a man was thrown from a scaffold by the violence of an explosion which took place when the man lowered a lighted candle into a tall cylinder used in connection with ammonia refrigeration by the absorption process.

Tendency of the Gas to Rise.

When a pipe that conveys ammonia bursts, anybody who happens to be near it should keep his head as low as possible while effecting his escape, because the gas being only half as heavy as air naturally rises as soon as it is liberated into the air; if a man stood erect he might possibly be overcome by the gas, while if he stooped he would, in a great many cases, escape without experiencing any bad effects.

Solubility in Water.

As ammonia is exceedingly soluble in water (so much so that 1 part of water will at 60° Fahr. absorb about 800 parts of the gas) the latter should be used to "kill" the gas in the event of any considerable quantity of strong ammonia solution being spilt. Also, in the case of a man going to the rescue of anybody who is overcome by the gas, he should first take the precaution of placing a piece of waste or rag soaked with water over his nose and mouth before entering the atmosphere that is impregnated with ammonia.

ACTION ON COPPER.

No part of an ammonia apparatus with which the ammonia is liable to come directly in contact must be constructed of copper or any of its alloys, such as brass, bronze, etc., as the parts containing that metal will be rapidly eaten away.

26° AMMONIA.

Commercial liquid ammonia, commonly known as "spirits of hartshorn," is a solution of ammonia gas in water. In the wholesale trade it is sold in large iron drums, and as its usual strength is 26° Beaumé, it is known as "26° ammonia."

ANHYDROUS AMMONIA.

The other commercial preparation of ammonia is liquid anhydrous ammonia, and it must not be confounded with the ordinary liquid 26° ammonia. The difference between the two is that the liquid anhydrous (from the Greek vdōr — meaning without water) ammonia is the pure, dry, ammonia gas compressed to a liquid, while the 26° ammonia, as we have already seen, is a solution of the gas in water.

CHAPTER III.

GENERAL ARRANGEMENT.

Users of ammonia refrigerating machines arrange their plant in a manner that best suits their special requirements or accommodations; but wherever it is practicable the whole of the plant should be as compact as possible, so that the possibility of loss of refrigerating effect due to the absorption of heat by long connections from the surrounding atmosphere may be reduced to a minimum.

Figs. 3 and 4 show the principal parts of an ammonia plant, arranged so that the following explanation can be easily followed and understood:

Description of the Plant.

When the plant is in working order the liquid anhydrous ammonia is contained in the receiver, E, and the bottom two or three coils of the condenser; and being under a gauge pressure of, say, 120 lbs., it flows through the pipe F and the manifold G to the expansion valves, H. Passing through the expansion valves, the ammonia traverses a series of pipes or coils which are surrounded by brine in the refrigerator, I, and terminate in the manifold K, that leads to the suction of the compressor, A. The suction of the compressor maintains a gauge pressure of, say, 28 lbs. in these series of pipes, and thereby relieves the ammonia of its high pressure as soon as it passes the expansion valves. Directly the liquid anhydrous ammonia experiences this relief of pressure it commences

28   *Theoretical and Practical*

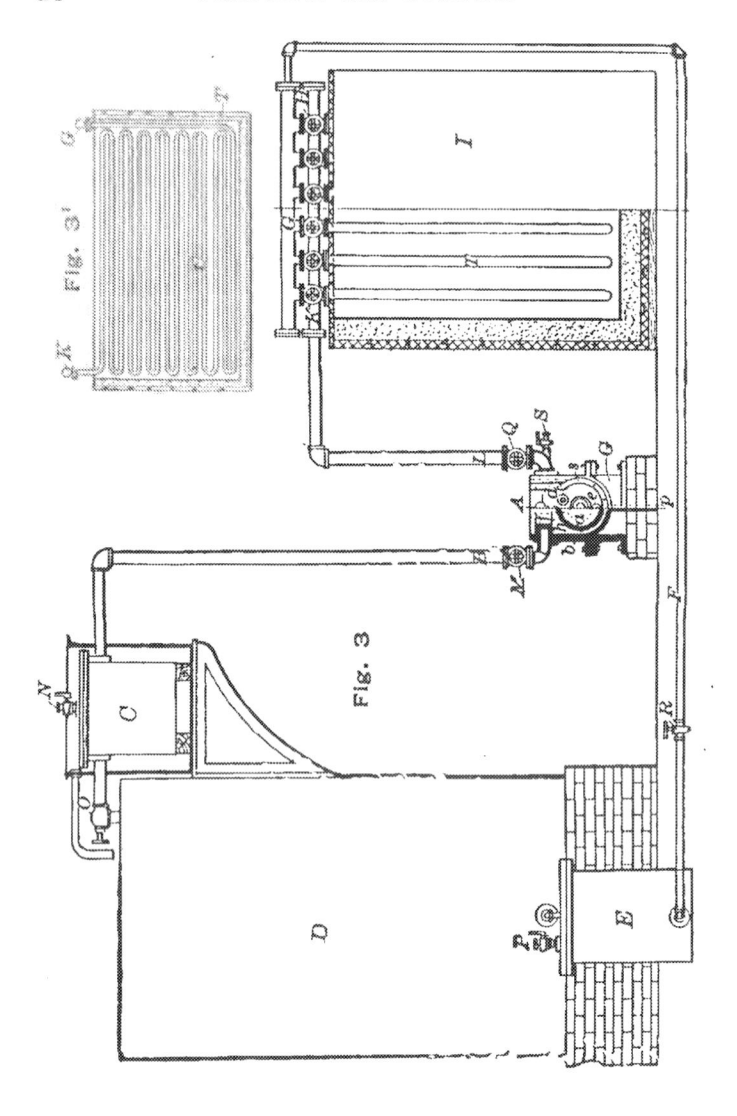

Fig. 3¹

Fig. 3

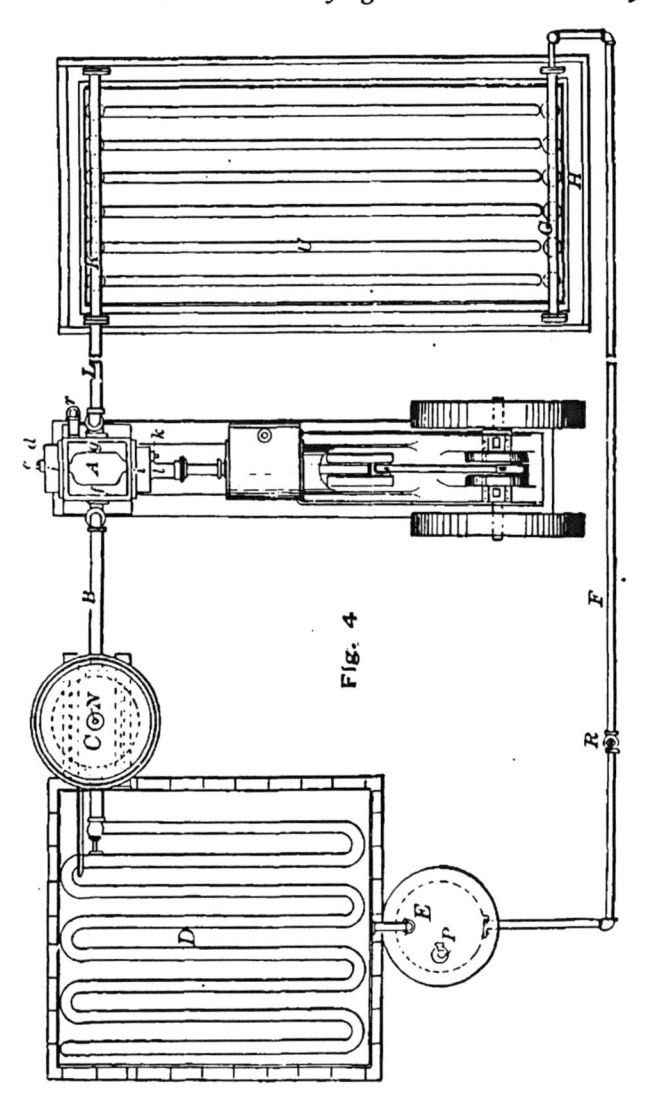

Ammonia Refrigeration.

Fig. 4

to boil, or vaporize, and in so doing it extracts heat from the brine, which latter could be cooled down to the boiling point of the ammonia due to a suction pressure of 28 lbs., namely, to 14° Fahr. By the time the ammonia reaches the manifold K it has been entirely vaporized, and therefore passes off in the gaseous state, and entering the compressor by the pipe L it is compressed and then discharged through the pipe B into the separator, C, where any of the oil (used for lubricating the compressor) or other foreign matters that are mechanically carried forward by the gas are separated, and the gas then enters the condenser, D, where it is again liquefied and, running down into the receiver, E, recommences the above-described movements.

Construction Details—The Compressor.

Owing to the heat that is generated during the compression of ammonia gas it is necessary that the compressor shall be surrounded, or jacketed, with water, so as to prevent the overheating of the cylinder, etc., and undue abrasion of the rubbing surfaces. The horizontal type of compressor is usually jacketed from end to end, but the heads are not artificially cooled.

A, Fig. 3, is a half-sectional end view of a horizontal compressor. The cylinder, a, and jacket, b, together with the gas passages, f and g, in Fig. 4, are cast in one piece, which is bolted to the engine frame, G. The passage g supplies the two suction valves, d and k, while the discharge valves, e and l, connect with the passage f. The jacket is supplied with water by the pipe p, the water filling up the space h and overflowing through r. The cylinder heads, $i\ i$, which contain the valves, ports and passages leading to f and g, are held in place by the bolts, s.

In the vertical type of compressor the water-jacket is built so that the water not only surrounds the compressor cylinder but also entirely submerges the cylinder head and its valves. The relative efficiency of the two types of compressors will be compared under the heading "Indicator Diagrams."

STUFFING-BOXES.

One of the principal sources of loss of ammonia in a refrigerating plant is in the stuffing-boxes of the compressor. The stuffing-boxes in some of the vertical types of compressors are packed with lead or babbitt-metal rings cut with a bevel, so that when they are subjected to pressure every alternate one hugs the piston-rod, while the others are pressed tightly against the inner surface of the stuffing-box, thus forming a tight yet smooth working packing. In the vertical compressor, which is only single-acting, the pressure on the packing does not exceed 28 lbs. on the square inch, while with the horizontal compressor, which is double-acting, the pressure may reach and even exceed 165 lbs., according to the temperature of the condensing water. For this reason it is necessary that the packing for stuffing-boxes in a horizontal compressor stuffing-box shall be deep. The depth is usually 12 inches, and the annular space between the piston-rod and the inside of the box is about ⅝ of an inch. It requires a considerable amount of attention which is more or less proportional to the condensing pressure, but more especially to the kind of packing that is used, and it is with a sense of the benefit that the user will derive that "Common Sense," "Garlock's," and "Selden's" packings are recommended as being specially suitable (if used conjointly) for horizontal compressor stuffing-boxes. The most satisfactory way to employ this combination packing is to, first of all, pack the stuffing-box to a depth of 5 to 5½ inches with Common Sense packing; then, having placed the perforated ring in position, half fill the rest of the box with Garlock's packing and finish off with Selden's packing.

The packing should be driven tightly home, piece by piece, and then the gland should be screwed on only hand-tight, so as to allow the packing room to expand and fill the spaces without undue pressure. If the packing is forced into the stuffing-box by means of the gland, and is not allowed room to expand, it will last but a very short time, and give trouble as long as it does last.

Special Lubrication.

The hot ammonia gas under high pressure will cut through the best packing in a very short time if a liberal supply of oil is not forced into the stuffing-box at intervals of an hour or so. To effect the thorough lubrication of the packing it is necessary that a hole shall be tapped in the centre (longitudinally) of the stuffing-box, which is then connected by a $\frac{1}{4}$-inch pipe with a small hand force-pump. The packing is divided into two portions by a perforated iron ring, which ring is directly opposite the above-mentioned hole, so that when the oil is delivered by the pump it is distributed through the perforations to the packing on either side of the ring.

Oil for Lubrication.

On no account must any animal or vegetable oils be used for lubricating the compressor, because as soon as any of these oils come in contact with the ammonia they will form soaps that will give endless trouble and annoyance. Nothing but a mineral oil of high viscosity and guaranteed purity should be used.

Clearance Space, Etc.

It is very essential that there shall be no unnecessary spaces, such as screw-slots, deep ports, etc., on the inside of the compressor cylinder, and the clearance space between the piston and cylinder head should not exceed 1-32d to 3-64ths of an inch. If attention is not paid to these particulars too much gas will remain in the cylinder after the piston has completed its stroke, and the re-expansion of this clearance-space gas as the piston recedes will greatly diminish the working capacity of the cylinder.

36 *Theoretical and Practical*

SUCTION AND DISCHARGE VALVES.

The suction and discharge ports are closed by poppet valves. The discharge valve, Fig. 5, screws into the outside of the cylinder head, and the spring, *a*, presses the valve

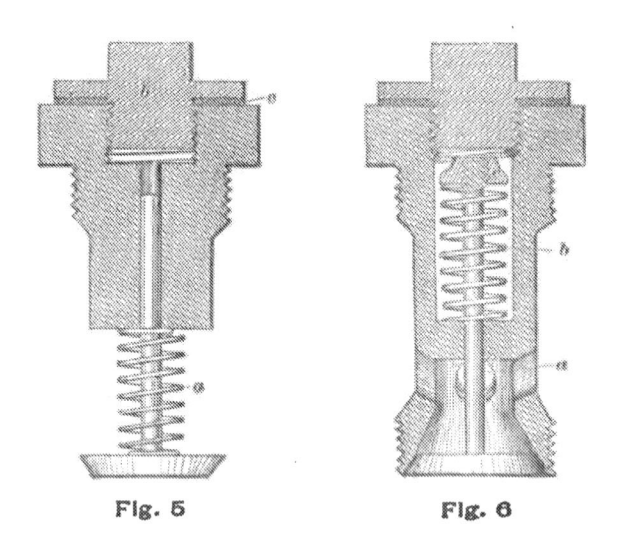

Fig. 5   Fig. 6

against the seat on the inside of the head. The suction-valve, Fig. 6, screws into both the outside and inside of the cylinder head, and the gas in G, Figs. 3 and 4, passes in through the holes, *a*, in its passage to the cylinder. The spring, *b*, is held in its place by the nut, *c*.

EFFECT OF EXCESSIVE VALVE-LIFT.

The lift of the valves is of very great importance, as it materially affects the refrigerating effect of a machine. If the lift is too great the valve will not act with sufficient quickness, and especially is this so in the case of high-speed compressors, in which an additional valve-lift of $\frac{1}{8}$ of an inch will cause a diminution of one ton refrigerating effect in 24 hours.

REGULATION OF VALVE-LIFT.

The lift of the discharge valve is regulated by the plug, *b*, against which the valve-stem strikes, the distance between the striking surfaces being regulated by the thickness of gasket, *c*. In the case of the suction-valve, the lift is regulated by means of an iron sleeve around the valve-stem against which the nut, c, strikes when the valve opens.

CHAPTER IV.

THE SEPARATOR.

OWING to the large volume of oil that is, or should be, used for lubricating the stuffing-box of the compressor, it is evident that a considerable quantity of it must pass into the cylinder and be carried through the discharge valves by the ammonia gas. If this oil were allowed to pass into the condenser it would soon find its way into the rest of the apparatus, and would cause trouble by choking up the expansion valves, etc.; therefore, with a view to obviating this annoyance, a separator is interposed between the compressor and condenser. The usual form of separator is an iron cylinder about 18 inches in diameter and from 18 to 36 inches high. The ammonia gas enters by a connection on one side and leaves by a connection on the opposite side. The connections are usually 3 or 4 inches from the top, and the gas coming in contact with the side of the cylinder is freed of the most of its oil and passes on to the condenser, while the oil falls to the bottom of the separator. This and most other forms of separators are very imperfect, for the reason that they are not supplied with sufficient contact-surface and are not kept sufficiently cool. The gas when it passes through the separator is at a high temperature, say 200° Fahr., and consequently the oil held in suspension is exceedingly limped and light in weight, and has not any great tendency to separate from the gas. The author would, therefore, advise the construction of a separator on the principle shown in Fig. 7. The cast-iron cylinder, A, with its inlet, E, and outlet, F, opposite one another, has its cover, B, and contact plates, C, cast in one piece, and these are arranged so that when the gas impinges on them it is

40 *Theoretical and Practical*

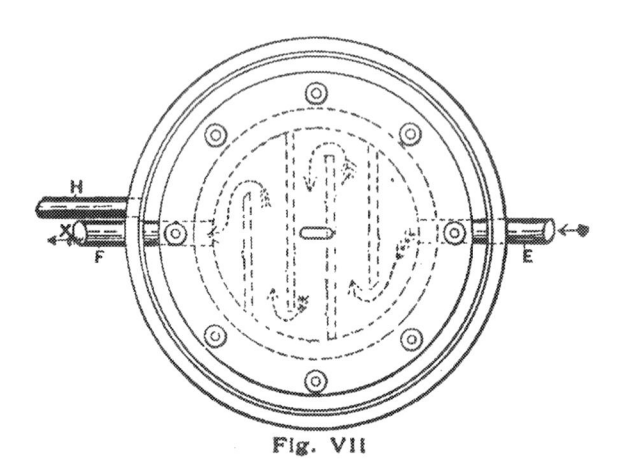

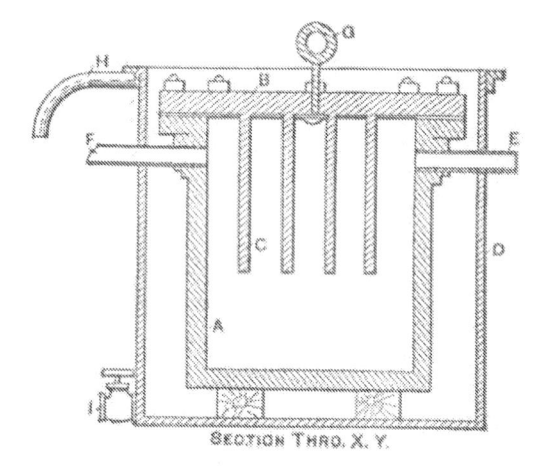

Fig. VII

Section thro. X.Y.

distributed over a large surface and is forced against the side of the cylinder in its zigzag passage from E to F. The oil in striking against these division plates will separate from the gas far more readily than if it meets with no obstruction, but even with the aid of the contact plates the separator will not effect a perfect separation unless the oil is rendered more viscous so as to increase its tendency to adhere to the plates, etc. This can be easily accomplished by making use of the water-jacket, D, which will keep the separator cold enough to make the oil separate and fall to the bottom. The bottom of the separator may be connected with the compressor so that the separated oil may be used over again; but this connection is of little or no use with double-acting compressors, because pieces of packing, etc., that find their way from the stuffing-box into the compressor and thence into the separator will soon choke it up. The separator should be periodically cleaned, the cover, B, and plates, C, being raised by the ring, G, after the water has been run off from the jacket by the cock, I. On no account must the inlet to the separator look down, because the gas will then impinge on the oil lying in the bottom, and will be likely to become more contaminated with, rather than freed of, the oil.

The Condenser.

The shape of the condenser tank affects the efficiency of the condenser to some extent: it should be deep and narrow rather than long and shallow, so that there may be as great a distance as possible between the more or less warm water on the surface and the cold water that is admitted at the bottom. Another important point is to see that the water is properly distributed when it enters the bottom of the condenser, and not allowed to all run in at one point, as in the case of a discharge through an open-end pipe.

Condenser-Worm.

The condenser-worm or piping through which the ammonia passes should consist of about one-third of 2-inch, one-third of 1½-inch, and one-third of 1-inch pipe. This gradual decrease in the size of the pipe will give far less "excessive" condensing pressure than when the gas passes from a manifold into a series of three or four separate one-inch worms. The friction of the gas in passing through a 2-inch pipe is less than when the gas passes through a number of pipes whose aggregate areas are equal to a 2-inch pipe. Another point is, it is quite unnecessary to have the same cross-sectional area for the exit as for the inlet pipe, because the volume of the liquid anhydrous ammonia passing through the exit is only about 1-75th of the volume of the gas that passes through the inlet pipe.

Receiver.

The receiver should be capable of holding 4 lbs. of liquid anhydrous ammonia for every 24-hour-ton maximum capacity of the machine. That is to say, if the machine has a maximum capacity of 65 tons of ice in 24 hours, the receiver should be capable of holding 65 × 4 = 260 lbs. of liquid anhydrous ammonia.

Refrigerator or Brine Tank.

The arrangement of the piping in the refrigerator is different from that in the condenser. By referring to Fig. 3 it will be seen that the liquid ammonia entering the series of piping at the manifold G descends by the vertical pipes, T, and then passes upward through the coils, U, before it is taken into the suction manifold K. The object of arranging the piping in this way is to insure the thorough vaporization of the liquid ammonia when the brine has become cooled down to a point near to the boiling point of the ammonia due to any given suction pressure, and the vaporization is thoroughly effected because any liquid ammonia that does not vaporize will not pass upwards, and therefore the gaseous or vaporized ammonia has to bubble through it, and the liquid thereby absorbs sufficient heat from the gaseous ammonia to effect the vaporization of the whole. If the liquid ammonia passed in at the higher and out at the lower extremity, as in the case of an ordinary condenser-worm, a large quantity of the ammonia would pass through in the liquid form, as the warmer, or gaseous portion, would not be brought so intimately in contact with it. The refrigerator should be thoroughly insulated, and for this purpose it should be surrounded by a wooden jacket so that there is a space of about 3 to 6 inches between the refrigerator and the inside of the jacket, and this space should be filled with mineral-wool, charcoal, sawdust, or any other good nonconductor.

SIZE OF PIPE AND AREA OF COOLING SURFACE.

The size of pipe and total cooling surface exposed to the brine very materially affect the economical running of a refrigerating plant, and practical results have demonstrated without doubt that coils, or worms, made of 2-inch pipe are far more economical in regard to the use of steam, etc., than 1-inch pipe. The total length of piping in contact with the brine should be sufficient to give a mean cooling surface of 50 to 55 square feet per 24-hour-ton maximum capacity.

EXPANSION VALVES.

The expansion valves are of the spindle type as shown in Fig. 8, and should be made of the best quality of cast-iron.

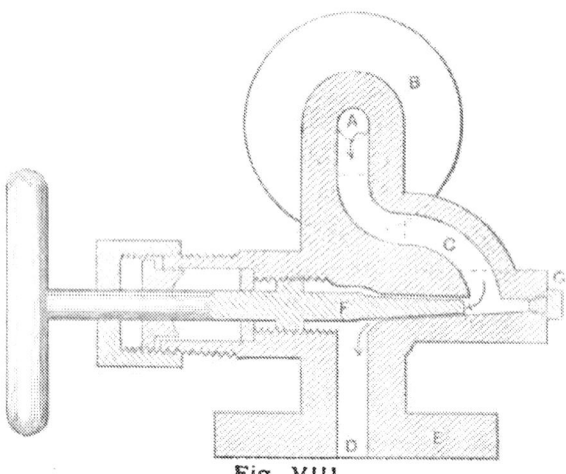

Fig. VIII

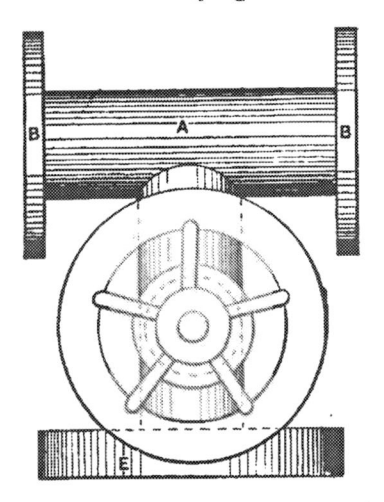

A = Manifold when number of valves are connected by flanges, B, B.

C and D = Inlet and Outlet Passages.

E = Flange connecting valve with coil in refrigerator.

F = Needle-Valve.

G = Plug to simplify cleaning passages in case of stoppage.

Working Details.—Charging the Plant with Ammonia.

In order to charge a new or at any rate an empty plant with ammonia it is first of all necessary to expel the air. This is done by opening all the valves and cocks with the exception of O, P, and S, which latter are tightly closed, and allowing the compressor to exhaust the air from D, E, F, G, I, K, and L, and discharge it through the open cock N, until the combination vacuum-pressure gauge connected to the suction. of the compressor shows that the engine is not capable of exhausting the apparatus any further; the cock N and valve H are then closed and the valve O opened. The drum of anhydrous ammonia (if an anhydrous ammonia generating apparatus is not included in the plant) is now connected with the cock S, which latter is then opened to allow the compressor to transfer the ammonia from the drum. When the plant is charged the cock S is closed and the valves H are then opened sufficiently to allow the compressor to maintain the suction pressure corresponding to the required brine temperature, which will be alluded to later.

CHAPTER V.

AMMONIA TO BE GRADUALLY CHARGED.

THE plant should not be charged with more than 60 per cent. of its full complement of ammonia at its first charging because it is impossible to exhaust the whole of the air from the plant by means of the compressor, and the only way to get entirely rid of the air is by displacement. This is effected by very cautiously opening the cock P once or twice a day and allowing the air to escape, at the same time taking every precaution to prevent undue loss of ammonia. After the air has been displaced a fresh quantity of ammonia is pumped into the plant in the manner above described, and the next day the same operation is gone through again, until at the end of, say, six days, the full complement of ammonia has been charged. In this manner the whole of the air is effectually expelled with but a slight loss of ammonia. An experienced man can easily tell from the general conditions and working of the plant when sufficient ammonia has been charged; but as the uninitiated might experience some difficulty in ascertaining whether the plant was sufficiently charged, the following method has been formulated for calculating the quantity of ammonia that constitutes a full charge.

Suppose the maximum capacity of plant is 65 tons of ice per 24 hours, and that the sizes of the different parts are as follows:

		Diam.	Length.	Capacity. Cubic Ft.
Connection from Compressor to Separator.	B	2 in.	10 ft.	
Separator	C	24 "	2 "	41.1
Condenser-Worm. { Containing Ammonia as gas.	D¹	1½ "	2800 "	
Condenser-Worm. { Containing Ammonia as liquid.	D²	1½ "	700 "	
Receiver	E	24 "	3 "	
Connection from Receiver to Refrigerator	F	1 "	30 "	18.3
Manifold for Expansion Valves	G	2 "	6 "	
Refrigerating Piping	T & U	1½ "	6000 "	
Connection from Refrigerator to Compressor	K & L	2 "	10 "	100.3

The parts B, C, and D[1] will contain ammonia in the gaseous state at a gauge pressure of, say, 120 lbs. and average temperature of 80° Fahr.

The parts D[2], E, F, and G will contain liquid anhydrous ammonia.

The parts T, U, K, and L will contain gaseous ammonia at a gauge pressure of 28 lbs. and an average temperature of 15° Fahr.

TABLE I.

Gauge Pressure.	TEMPERATURE OF GAS.						
	66°	74°	80°	84°	90°	95°
	Volume of 1 lb. of Gas in Cubic Feet.					
80	3.470					
85	3.292					
90	3.131					
95		3.035				
100		2.900				
105		2.785				
110			2.695			
115			2.590			
120			2.490			
125				2.418		
130				2.333		
135				2.252		
140					2.204	
145					2.134	
150						2.088
155						2.037

From Tables I. and V. it will be seen that the volumes of the ammonia gases at the above *pressures and temperatures* of 120 lbs. and 80° Fahr. and 28 lbs. and 15° Fahr. are respectively 2.490 and 10.763 cubic feet per pound of ammonia; therefore the amount of ammonia required to charge the plant is:

B, C, and D¹.........	= (41.1 ÷ 2.49)	16½ lbs.
D², E, F, and G.......	= (18.3 × 38.66*)	707½ "
T, U, K, and L........	= (100.3 ÷ 10.763)	9⅜ "

Total, 733⅜ lbs.

JACKET-WATER FOR COMPRESSOR.

The amount of jacket-water necessary for the compressor varies according to the condensing pressure. With a low condensing pressure—say 90 to 105 lbs. gauge pressure—10 to 15 gallons of water per hour per 24-hour ton refrigerating effect will usually be found ample, but when the condensing pressure reaches, say, 140 to 150 lbs., the amount of water will have to be increased to about

---

* Weight of a cubic foot of liquid anhydrous ammonia.

45 to 50 gallons per hour per 24-hour ton refrigerating effect.

JACKET-WATER FOR SEPARATOR.

The amount of water used in the separator jacket should be as large as possible, and so that the water may not be wasted or become expensive, the overflow-pipe, H, should be continued down midway into the condenser, where the water should be distributed and used along with the condensing water that is admitted at the bottom of the condenser.

CONDENSING WATER.

As the pressure against which the compressor has to work is regulated almost entirely by the temperature of the condensed ammonia, it is obvious that the lower the temperature of the condensed ammonia, the greater the saving in the wear and tear of the engine, in the use of steam and consequently the consumption of coal, will be. The quantity and the temperature of the condensing water are, therefore, points that need careful consideration. The manufacturer who has to use the city water-supply for condensing purposes can not, under ordinary circumstances, economically cool the ammonia to a lower temperature than 55° to 60° Fahr. during the winter months, and 65° to 75° Fahr. during the summer months, because, should he increase his supply of water sufficiently to reduce the temperature of the ammonia, say 10° below the above figures, he would at once incur an extra expense that would not be warranted by the resulting increase in the refrigerating efficiency of the plant. This increased expenditure can, however, be overcome if the following plan is adopted:

Lessening the Cost for Condensing Water.

Instead of supplying the steam-boilers in the establishment with the whole of their water direct from the main, the author advises arrangements being made to draw the boiler water-supply from the overflow of the ammonia condenser, then making up the deficiency from that source by drawing from the main. This method of working would be beneficial in every respect, because in the first place, the water in passing through the condenser will receive a certain amount of heat which is distinctly an advantage, as boiler-water is, or should be, heated before entering the boiler. Secondly, if the whole or a part of the water required for the boilers is taken from the ammonia-condenser overflow, the cost of condensing the ammonia is practically reduced to *nil*, because the boilers have to be supplied with water, and the fact that that necessary supply has been previously used for condensing purposes in no way increases the cost after the first cost of putting up the system of piping for conveying the water has been paid for. Thirdly, the effect of the use of a superabundance of condensing water will be a reduction of, at least, 30 to 40 lbs. per square inch in the condensing pressure and a corresponding saving in steam.

Quantity of Condensing Water Necessary.

If the temperature of the water supplied to the condenser is 55° to 60° Fahr., and the temperature of the overflow or outlet water is 85° to 90° Fahr., the quantity of water that will be required will be about 0.9 gallons per minute per 24-hour ton of ice; but if the temperature of the overflow were only 70° to 75° Fahr. (the inlet temperature being 55° to 60°), the quantity of water that would be necessary would be about 2½ gallons per minute per 24-hour ton of ice. This reduction of fifteen degrees in the temperature of the overflow means a reduction of 30 to 40 lbs. in the condensing pressure, and if the ammonia leaves the condenser at the temperature of the inlet water, a minimum condensing pressure and large saving in steam will result.

Loss Due to Heating of Condensed Ammonia.

One very weak point and very surprising oversight in the management of a great number of refrigerating plants is the fact that, although manufacturers often go to a deal of expense in order to condense and cool the ammonia to the lowest possible temperature, they entirely ignore the importance of making arrangements to maintain that low temperature until the ammonia reaches the refrigerator. The receiver, and a considerable length, if not the whole, of the piping through which the anhydrous ammonia has to pass on its way to the refrigerator are, as a rule, situated in the engine-room—which is not usually the coolest of places—and the temperature of the ammonia is consequently often raised 5, 10, 15, or even 20 degrees (above the temperature at which it left the condenser) before it reaches the refrigerator; and as these 5 to 20 degrees gain in temperature mean a loss of from $\frac{1}{4}$ to $1\frac{1}{4}$ ton refrigerating effect per 24 hours, on a 65-ton machine, it seems as though it would be advantageous to have the receiver and piping covered with a cheap non-conducting material, so as to take full advantage of the benefits resulting from a liberal water-supply to the condenser, and thus prevent an unnecessary waste.

Loss Due.

It might be advisable to here refer to another source of needless loss which has even a greater effect on the refrigerating efficiency of a machine than the case just considered.

Superheating Ammonia Gas.

It is the loss incurred by the ammonia gas absorbing heat in the transit from the refrigerator to the compressor. Some people argue that it is absurd to go to any expense for the purpose of preventing that gas from absorbing heat, as it is heated up, any way, as soon as it enters the compressor. Others, again, consider that any heat absorbed by the gas simply means that a few more thermal units will have to be extracted from the gas when it passes into the condenser. If these people would just take time to think, they would at once see that the higher the temperature of the gas is before it enters the compressor the greater the volume of a given weight must be, and therefore the compressor, although circulating or pumping the same volume, will not circulate so great a weight; and as the refrigerating efficiency of a machine is proportional to the weight of ammonia circulated, it is obvious that the higher the temperature of the gas before it enters the compressor, the smaller the refrigerating efficiency of the machine will be, the suction pressure being the same in both cases. The effect of covering the ammonia pipes is more particularly dealt with under the heading "Directions for Determining Refrigerating Efficiency."

CHAPTER VI.

EXCESS CONDENSING PRESSURE.

THE condensing pressure, when the apparatus is working, is always greater than the theoretical. This "excess" pressure is due almost entirely to the confining of the highly heated gaseous ammonia in the more or less limited space of the coils of the condenser, and varies greatly according to circumstances. When running at a low suction pressure, say atmospheric pressure, the excess condensing pressure should not be over 5 to 10 lbs., but when running with a suction-gauge pressure of 20 to 28 lbs. the excess pressure will vary from 40 to 60 lbs.

Cause of Variation in Excess Pressures.

The reason why there is such a large variation in the excess pressure is obvious: with 28 lbs. suction-gauge pressure, the compressor is pumping a three times greater weight of gas than it would pump if the gas were under only an atmospheric pressure, and therefore the condenser is crowded to a greater extent in the former than in the latter case. It may be argued that if the compressor is forcing into the condenser a three times greater weight of ammonia in one case than in another, the condenser at the same time will be relieved by the expansion valves of a three times greater weight of liquid ammonia, and one will thus counterbalance the other. It is, of course, true that the weight of liquid ammonia passing the expansion valves will be the same as the weight of ammonia gas entering the condenser from the compressor; but as the volume of a given weight of the gas at condensing temperature and pressure is about 75 times greater than the volume of the same weight of liquid ammonia, it is plain that if instead of pumping in 75 volumes of gas into the condenser we increase the amount three times, or to 225 volumes, the increased delivery from the condenser (by means of the expansion valves) of only two volumes is insignificant in comparison with the increased receipt from the compressor, and therefore the increase of excess condensing pressure is what might naturally be expected to accompany increased suction pressure.

Other Conditions that Affect Excess Pressure.

No table of the excess condensing pressures for various suction pressures would be of any practical use, because different makes of refrigerating plants give different results. The high speed (140 revolutions per minute) horizontal compressor invariably gives a greater excess pressure than the vertical compressor, which only has a speed of from 40 to 60 revolutions per minute. The method of connecting the condenser piping also affects the excess pressure considerably, and if four separate one-inch pipes, or worms, connected by manifolds are used, the excess pressure will be greater than if one continuous worm (starting at the top with two-inch piping and reducing to one-inch, as recommended in previous pages) is used. Also, the higher the condensing pressure due to the temperature of the condensing water the greater the excess pressure will be.

Use of Condensing Pressure in Determining Loss of Ammonia by Leakage.

As the condensing pressure is one of the principal means by which the engineer can tell when the loss of ammonia by leakage has amounted to such a quantity as to render the replenishing of the plant advisable, it is very necessary that the man in charge, if inexperienced, should record in a book the temperature of the condensed ammonia at its point of exit from the condenser, and the suction and condensing pressures, every two or three hours. If these figures are thoroughly memorized and the engineer started with a plant that was fully charged with ammonia he ought to be able, at the end of a month or two, to tell by looking at the suction-pressure gauge, and the temperature of the condensed ammonia whether the condensing pressure was what it should be. For example, suppose the plant has been running for two or three months with an average condensing temperature of 60° Fahr., condensing pressure of 120 lbs. and suction pressure of 25 lbs., and that during the next three months the condensing pressure gradually fell to 115 lbs., while the condensing temperature and suction pressure were still 60° Fahr. and 25 lbs. respectively; it would be plain that neither the condensing temperature nor the suction pressure could account for this falling off in the condensing pressure because they have not altered, and therefore it is obvious that the quantity of ammonia can alone account for this alteration. The diminution in the condensing pressure caused by loss or leakage of ammonia is due to the increased condenser space resulting from the leakage, thereby allowing the gas a greater length of worm in which to condense and assume the liquid form, thus lessening the "crowding" of the hot compressed gas.

When the condensing pressure falls off 5 or 10 lbs. the plant should be re-charged with sufficient ammonia to restore the normal condensing pressure.

Cooling Directly by Ammonia.

It is very seldom that ammonia can be used directly for freezing purposes, and in nearly all cases it is used indirectly with brine as a medium. The greatest drawback to using ammonia directly is the liability of ammonia to leak through the fittings, joints, etc., and as meats or other provisions would be rendered valueless as far as the market is concerned by such a leakage, it would be exceedingly risky and injudicious to cool a warehouse directly by ammonia if the only object for so doing was to save the cost of the brine portion of the plant. But in buildings where a slight smell of ammonia would not result in any pecuniary loss—other than the value of the escaping ammonia, which latter if properly looked after will be exceedingly small—it would certainly be advisable to cool directly by ammonia. In this case the expansion valves would be in the building to be cooled, and the ammonia would be expanded in a system of piping hung up on the walls or otherwise conveniently arranged. This method of working is decidedly the most economical, as it does away with the necessity of a refrigerator and its long series of piping, the brine pumps and the steam required to run them, the brine piping (4 to 5 inches in diameter) conveying the brine between the pumps, building to be cooled, and the refrigerator, and all the numerous fittings and valves in connection therewith.

BRINE.

Brine is a solution of either common salt (chloride of sodium), chloride of calcium, or chloride of magnesium in water. Brine made of chloride of magnesium is undesirable, as it is liable to contain free acid, which above all other things is most objectionable, owing to its action on metals; whereas common salt, or the "commercial fused" chloride of calcium, are both free from acid. Salt is usually sold by the bag, each bag containing about 200 lbs. and costing about 70c., or $7.00 per ton. Commercial fused chloride of calcium is sold in iron drums, holding about 600 lbs. each, and costs about $16.00 per ton. Cheap common salt, such as may be obtained for 40 to 50 cents per bag, should not be used, as it will be expensive in the long run, and nothing but the purest and best salt should be bought. Common salt for brine making should not contain more than 0.05 per cent. of insoluble matter (calculated on the dry salt). The percentage of moisture is only of account when the salt is bought by weight instead of by the bag, but the percentage of insoluble matter is always of great importance, because, unless there are special facilities for filtering the brine before it enters the refrigerator or system of piping for cooling rooms, etc., it is obvious that if the percentage of insoluble matter is bulky, it will accumulate and eventually settle down in the bottom of the refrigerator and thereby reduce the efficiency of the apparatus by covering the piping, or it is liable to pass into the brine pumps, and from thence to the brine piping for cooling the rooms, where it is likely to lodge in fittings (return bends, elbows, etc.) and cause serious obstruction. The use of chloride of calcium does not do away with the inconvenience liable to be caused by the presence of insoluble matter, but for temperatures below $-7°$ Fahr. it is absolutely necessary that it should be used for the reason explained in paragraph on "Effect of Composition on Freezing Point."

Freezing Point of Brine.

Brines will only stand a certain degree of cold without freezing, and the temperature to which brine can be cooled before it will begin to freeze depends, firstly, on the composition of the brine, and secondly, on the strength of the solution.

Effect of Composition on Freezing Point.

In illustration of the effect that a change in the composition of the brine will have on the freezing point it is only necessary to state that whereas a solution of common salt can only be cooled to $-7°$ Fahr., a solution of chloride of calcium can be cooled to $-40°$ Fahr.

EFFECT OF STRENGTH ON FREEZING POINT.

In explaining the way in which the strength affects the freezing point of the solution a brine made of common salt will be considered. If a weak solution of common salt in water is gradually cooled, ice will begin to separate out at about $28°$ Fahr., and this separation of ice with a proportional concentration of the brine will continue till the temperature of $-7.5°$ Fahr. is reached. At this point the brine will contain 24.24 per cent. of salt, and if further cooled will solidify as a whole. If, on the other hand, a saturated solution (at $60°$ Fahr.) of salt is cooled, salt will separate out, and the brine will weaken until the same temperature and degree of concentration given above is reached, when the solution will become wholly solidified.

Suitableness of the Brine.

For all ordinary purposes, such as ice manufacture, etc., where it is highly improbable that a temperature below $-7°$ Fahr. will be needed, the author would strongly advise the use of a brine made of common salt. The cost is less than one-half of that of chloride of calcium, and it is far easier and more cleanly to handle, because chloride of calcium is highly deliquescent, and therefore a drum of it must be used as soon as opened, otherwise it will absorb so much moisture from the air that it will "run" and cause much annoyance—not to mention loss. As we have already seen, if the brine is either too weak or too strong, a separation will take place—in the former case of ice, and in the latter case of the chemical constituent. Now, if either of these separations occurs it will seriously affect the refrigerating efficiency of a plant, owing to the coating of the refrigerator coils or piping with a bad conducting material such as ice, salt, or chloride of calcium. It is therefore of the greatest importance that the gravity or strength of the brine should be carefully tried every day, and any variation due to evaporation or other causes should be corrected at once.

Making Brine.

The brine should be made in a separate vessel and not be transferred to the refrigerator until its strength has been carefully adjusted and the dirt, etc., allowed sufficient time to settle to the bottom. If the brine is to be made from salt, the water is first placed in the vessel and carefully measured, and then the requisite quantity of salt—namely, 266.81 lbs.* per 100 gallons of water—is thrown in and the whole stirred either mechanically or manually until the salt is dissolved. The strength of the brine should then be 22° Beaumé. In the case of chloride of calcium the strength can not be regulated to such a nicety as in the case of salt, because the

---

* These figures are for pure, dry salt, and therefore the percentage of moisture and insoluble matter contained in the salt used must be determined and allowed for.

material has to be placed in the vessel in more or less large lumps, and as these lumps dissolve comparatively slowly at the ordinary temperature it is necessary to boil the water with open steam. This operation, of course, increases the volume of the water first placed in the vessel, and as this increase is an uncertain quantity (according to the size of the lumps and therefore the length of time they take to dissolve) the strength has to be regulated entirely by the use of the hydrometer. It is wiser to make the solution too strong rather than too weak, as it takes less time to reduce the strength by adding water than it does to increase the strength by dissolving more of the chloride of calcium.

CHAPTER VII.

It is advisable to place only 6 gallons of water for every 100 lbs. of chloride of calcium in the vessel to start with, and as soon as the solution is effected cold water should be added, small quantities at a time, until the strength is reduced to 20° Beaumé.

Specific Heat of Brine.

According to Professor Denton,[*] the specific heat of brine made from common salt is as follows:

Strength.	Specific Heat.
20¼° Beaumé	0.818
21½° "	0.786

The author finds that the specific heat of brine of 22° Beaumé strength and made from American salt is 0.765.

Regulation of Brine Temperature.

In places where the refrigerating work is regular and the temperature of the brine returning to the refrigerator is not liable to vary many degrees, the regulation of the temperature of the outgoing brine is an easy matter; but where the return brine is sub-

---

[*] Transactions of the American Society of Mechanical Engineers, Vol. XII., page 384.

jected to large variations in temperature the regulation of the outgoing brine temperature requires a great deal of attention. In the former case the expansion valves are regulated so that the engine maintains a suction pressure equivalent to a boiling-point (of the anhydrous ammonia) of about 15° Fahr. lower than the brine temperature required. For instance, in ice-making a brine temperature of 25° Fahr. would be the most economical, and 15° lower than that, namely, 10° Fahr., would be the temperature at which the ammonia should boil. By referring to Table III. (page 116) it will be seen that a suction-gauge pressure of 23.85 lbs. is equivalent to an ammonia boiling-point of 10° Fahr., and therefore the expansion valves would need to be regulated so that the engine ran with a suction-gauge pressure of, say, 23¾ lbs. If a building has to be cooled and maintained at a temperature of zero, a brine temperature of about − 10° Fahr. will be necessary, and 15° lower than that (= − 25° Fahr.) will be the required boiling-point of the ammonia, and Table III. shows that a suction-gauge pressure of 1.47 lbs. corresponds to that boiling-point. In both these cases the expansion valves will need little or no attention after they have once been properly regulated; but it will now be shown that if we have a quantity of hot oil that has to be cooled a certain number of degrees Fahrenheit in a given length of time, it is necessary that the expansion valves shall be frequently attended to in order to obtain the desired results. For example:

50,000 lbs. of oil at a temperature of
100° Fahr. have to be cooled to
20° Fahr. or through
80 Fahrenheit degrees in
24 hours, and the specific heat of the oil is 0.750.

In this case the number of thermal units to be extracted from the oil are (50,000 × 80 × 0.750) 3,000,000. Now, if the compressor is capable of circulating 43,200 cubic feet of ammonia gas per 24 hours, and the expansion valves are regulated to give, at the commencement, a brine temperature of 15° Fahr., the refrigerating efficiency will be only 2,497,000 thermal units per 24 hours, and it will therefore take about 29½ hours to cool the oil to 20° Fahr. But if the expansion valves are regulated so that for the first six hours the brine temperature will be 32° Fahr. and during the next 12 hours 25° Fahr., and the remaining six hours 15° Fahr., the refrigerating efficiency will be, approximately:

First 6 hours	—	882,000	Thermal units.
Next 12 "	—	1,542,000	" "
Last 6 "	—	624,000	" "
Total,		3,048,000	Thermal units,

or 48,000 thermal units more than are theoretically required, and 551,000 thermal units more than could be extracted by starting with, and maintaining for 24 hours, the required final brine temperature of 15° Fahr. This great difference in the results is due to the simple fact that the refrigerating efficiency of a plant is proportional to the weight of anhydrous ammonia circulated, and therefore if a large weight of ammonia is circulated at the commencement, when the temperature of the oil is high, and that weight is gradually reduced as the oil becomes cooled, it is evident that the oil will be cooled quicker than if the smaller weight, or that necessary for the final temperature, is circulated throughout the whole of the operation. Of course, it would not be advisable to regulate the expansion valves so as to cause the three sudden drops in temperature as in the above example—where it was done for simplicity's sake—but the valves should rather be gradually closed, so that the minimum brine temperature required will be reached about six hours before the material that is being cooled will be required.

INDIRECT EFFECT OF CONDENSING WATER ON BRINE TEMPERATURE.

If the supply and temperature of the water used in the condenser is irregular the expansion valves will need constant attention (no matter what the nature of the refrigerating work may be), because any irregularities in the condensing water will cause changes in the condensing pressure. If the supply lessens in quantity the temperature of the condenser will, of course, rise and cause an increase of pressure. The natural result of increased pressure will be a larger delivery of ammonia forced through the expansion valves, and the suction pressure will in turn also be increased. It is therefore necessary to counterbalance increase of condensing pressure by a proportional closing down of the expansion valves, and decrease in the condensing pressure by opening the expansion valves.

CHAPTER VIII.

DIRECTIONS FOR DETERMINING REFRIGERATING EFFICIENCY.

BEFORE going into the details of determining the efficiency of a refrigerating plant it is necessary that one or two points in connection therewith should be explained.

EQUIVALENT OF A TON OF ICE.

The equivalent of a ton of ice is 284,000 British thermal units, or the amount of heat that would be necessary to convert a ton (2,000 lbs.) of ice at 32° Fahr., into a ton of water at 32° Fahr., or, conversely, it is the amount of heat that must be extracted from a ton of water at 32° Fahr. in order to convert it into a ton of ice at 32° Fahr.

COMPRESSOR MEASUREMENT OF AMMONIA CIRCULATED.

Professor Denton's determinations[*] show that when the ammonia gas enters the compressor it is heated by the walls of the latter and so rarefied as to cause the compressor full of gas to weigh upwards of 25 per cent. less than it would if the gas remained at the temperature of the entrance while the compressor filled.

---

[*] Transactions of the American Society of Mechanical Engineers, Vol. XII.

Loss in Well-Jacketed Compressors.

The make of machine with which Denton experimented was the Consolidated Ice Machine Company's, and the actual loss in the pumping efficiency of the compressors due to the above cause was 21.4 per cent. The compressors (including gas passages, valves, etc.) in this make of machine are exceptionally well arranged for receiving the fullest possible benefit from the jacket-water, and therefore the loss of pumping efficiency is reduced to a minimum. Where compressors are not so efficiently jacketed, the loss by superheating will vary from $21\frac{1}{2}$ to 25 per cent.

Loss in Double-Acting Compressors.

An allowance of 30 per cent. for loss by superheating is necessary in the case of double-acting compressors when the gas enters the compressor through the heads and the heads are not jacketed.

Before the efficiency of a plant can be determined it is necessary that the compressor should be fitted with an indicator, the engine and brine pumps with stroke counters, and that mercury wells should be placed at the following points, viz. :—

Distribution of Mercury Wells.

(1) On the discharge pipe, near its point of outlet from the compressor.

(2) On the ammonia discharge pipe from the condenser—immediately at its point of exit.

(3) In the ammonia supply manifold of the refrigerator.

(4) In the ammonia suction—or discharge—manifold of the refrigerator.

(5) In the ammonia suction pipe—immediately at its point of entry to the compressor.

(6) In the return brine pipe, just where it discharges into the refrigerator.

(7) In the brine discharge brine pipe from the refrigerator.

In cases where the pipes are horizontal and of sufficient diameter the mercury well should be constructed as in Fig. 9, in which A is the pipe, the temperature of the con-

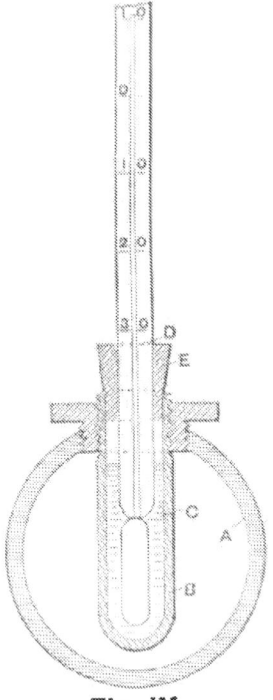

Fig. IX

tents of which is required; B is the mercury well, made of iron tubing and fitted in the pipe by means of a bushing. The mercury, C, fills the well about three-quarters full, and in it the thermometer, D, is held by the cork, E.

When the pipes are vertical, or of too small a diameter, the mercury well should be made as follows (Fig. 10):—

The wooden block, B, having a cavity, C, is carefully fitted to the pipe, A, and securely fastened in its place by the iron bands D, D. C is filled three-quarters full with mercury, and the thermometer, E, having been introduced and secured in its place by the cork, F, the whole is so wrapped in hairfelt as to entirely prevent any possibility of the atmosphere having any effect upon the temperature of the mercury.

The portion of the pipe with which the mercury comes in contact should be thoroughly scraped, so as to present a perfectly bright and clear surface, before the wooden block is fastened in its place.

The judicious application of a little soft putty to touching surface of the wood will make the joint between the wood and pipe perfectly tight and efficient.

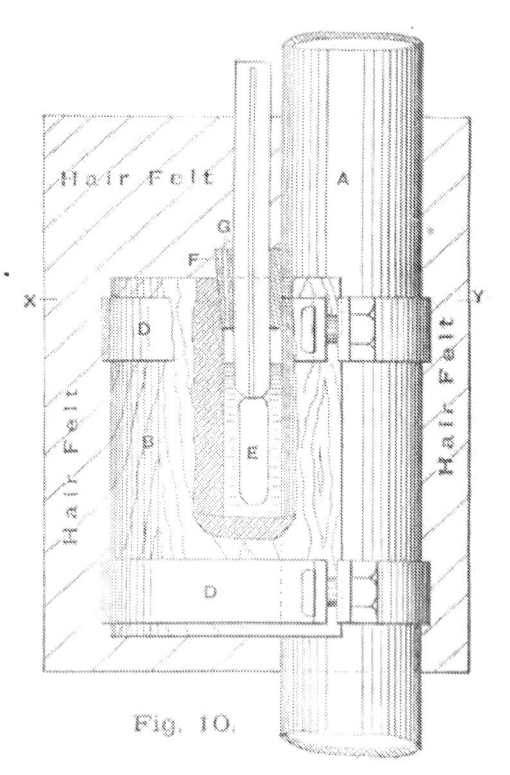

Fig. 10.

The most convenient form of thermometer is one with a cylindrical bulb $\tfrac{7}{8}$ to 1 inch long; the diameter of the thermometer should be about 5-16 to ⅜ of an inch. The graduations should start at a point 3 inches above the top of the bulb and should be

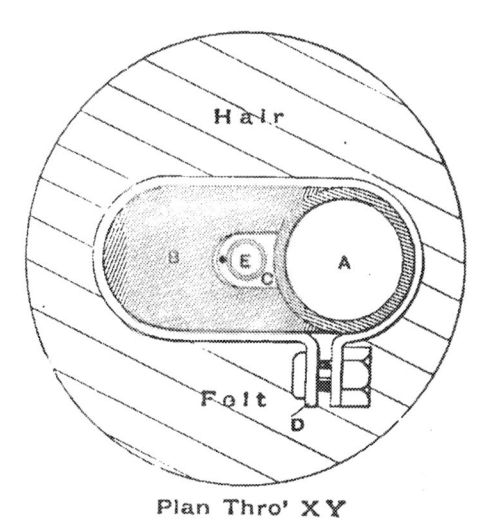

Plan Thro' XY

⅛ of an inch apart, and each graduation should represent one degree. With the use of such a thermometer a reading of one-tenth of a degree may be easily and accurately made.

Examination of Working Parts.

Having carefully examined the pistons and valves of the brine pumps and compressor, and verified the accuracy of the pressure gauges, a number of tabulated forms should be drawn up ready to receive the readings of the different instruments as they are taken.

Number of Readings to be Taken.

Where a plant is doing "steady temperature" work, such as cooling warehouses or making artificial ice, readings of all the different instruments need not be taken more than once every half-hour; but where the range in temperature of the material to be cooled is large, readings should be taken every quarter of an hour. Diagrams of the steam cylinder and the compressor should be taken every three hours.

CHAPTER IX.

DURATION OF TEST.

For steady work, the test should last for twelve hours, and in large range of temperature work the test should last for twenty-four hours, or, at any rate, until the final temperatures agree as closely as possible with those at the start.

INDICATOR DIAGRAMS.

In order to check the brine figures a very careful examination of the indicator diagrams of the compressor must be made, as it is only by the aid of these diagrams that an accurate computation of the volume of ammonia circulated can be made.

Fig. 11 represents the working of a double-acting horizontal compressor running at 140 revolutions per minute. The gauge pressure in the suction discharge pipes of the compressor when the diagram was taken were, respectively, 10 lbs. and 140 lbs. As the diagram shows that the suction pressure in the compressor was only 5 lbs. and the condensing pressure was 150 lbs., it is very evident, in the first place, that both the suction and discharge valves were too small and did

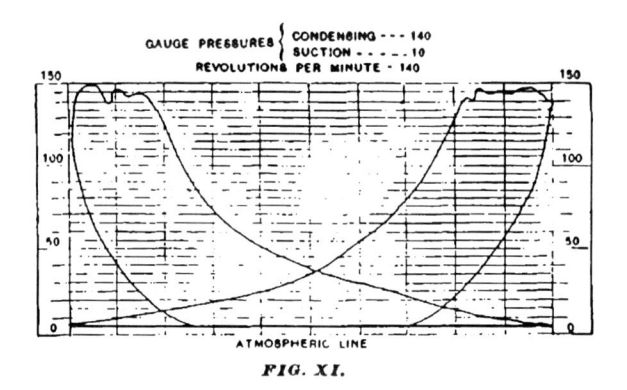

FIG. XI.

not admit of the free passage of the ammonia gas. Secondly, as the suction pressure in the compressor was only 5 lbs. the compressor was not pumping or circulating as much ammonia as the gauge pressure represented. This diagram also shows that the engine had performed 30 per cent. of its forward stroke and 25 per cent. of its return stroke before the pressure due to the re-expansion of the clearance space gas was reduced to the suction pressure—the pressure at which the valves would open. In this case the pumping capacity of the compressor was, therefore, only 72½ per

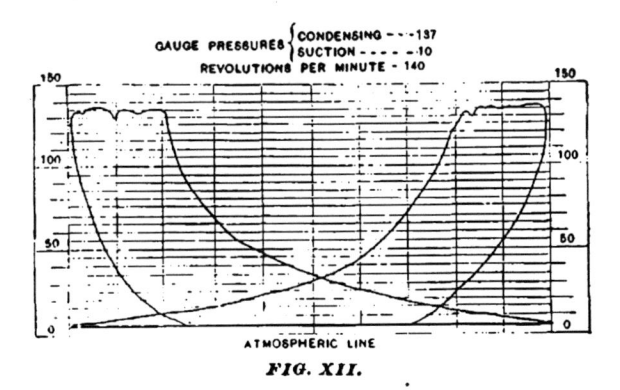

FIG. XII.

cent. of the piston displacement per revolution.

Fig. 12 represents the working of the same engine after the discharge valves had been enlarged. Although the engine was running at the same speed as before—140 revolutions per minute—the condensing pressure in the compressor was this time the same as indicated by the gauge on the discharge pipe, showing that the engine had no "excess" pressure to work against, and therefore a saving in steam was effected. The diagram again shows, however, that the suction valves were too small for a speed of 140 revolu-

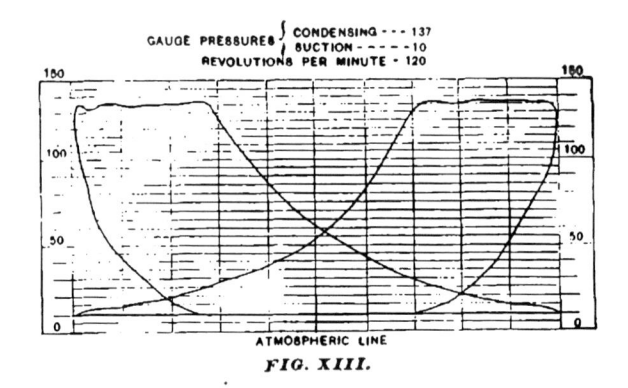

FIG. XIII.

tions per minute, and, also, that the pumping capacity of the compressor was only 72½ per cent. of the piston displacement.

Fig. 13 is the diagram taken from the same engine when running at the rate of only 120 revolutions per minute. From it we see that the suction valves of the compressor are designed for that rate of speed, and that the previous rates of 140 revolutions per minute were beyond the capacity of the valves.

Fig. 14 was a diagram taken from a compound single-acting vertical compressor running at 40 revolutions per minute, with a suction and condensing gauge pressure of,

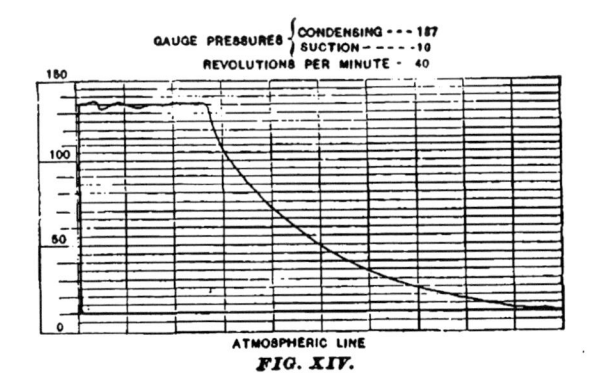

FIG. XIV.

respectively, 10 lbs. and 137 lbs. This diagram exhibits an almost perfectly square heel, the loss being only 1 per cent. of the piston displacement, and shows that the suction and discharge valves were of requisite size.

We will now see what these diagrams actually represent in pounds of ammonia circulated per 24 hours, and from those figures we will be better able to realize the importance of this portion of the subject.

For simplicity's sake we will suppose the temperature of the gas entering the compressor was 0° Fahr. in all four cases. The cubical displacement of the piston in the case of the horizontal compressor was 1.30 cubic feet per revolution, and in the case of vertical compressor 4.00 cubic feet per revolution.

140 revolutions per minute × 1.3 = 182 cubic feet per minute = 262,080 cubic feet per 24 hours.

The indicator diagram shows that 27.5 per cent. of this was lost owing to re-expansion of the gas, and we have seen under sub-heading "Loss in Double-acting Compressors," that 30 per cent. also has, in this case, to be deducted, and therefore the effectual displacement is ((262,080 − 27.5 per cent.) − 30 per cent.) = 133,005 cubic feet per 24 hours.

The suction pressure in the compressor was 5 lbs. (*i. e.*, 19.7, say, 19¾ lbs. absolute pressure). By Table V. (page 125) we see that 1 lb. of ammonia gas at 0° Fahr. and 19¾ lbs. absolute pressure = 14.828 cubic feet; therefore the effectual displacement of 133,005 cubic feet = 8,970 lbs. of ammonia circulated per 24 hours.

120 revolutions per minute × 1.3 = 156 cubic feet per minute = 224,640 cubic feet per 24 hours.

Taking 72.5 per cent. of this amount, and then deducting 30 per cent. of the remainder, we have an efficiency of 114,004 cubic feet per 24 hours.

The suction pressure in the compressor was 10 lbs. (= 24¾ lbs. absolute pressure). By Table V. (page 127) we see that 1 lb. of ammonia gas at 0° Fahr. and 24¾ lbs. absolute pressure = 11.794 cubic feet; therefore the effectual displacement of 114,004 cubic feet = 9,666 lbs. of ammonia circulated per 24 hours.

In the cases of diagrams 11 and 12, where the engine was running at a speed of 140 revolutions per minute, the pounds of ammonia circulated were only 8,970 as against

9,666 when the engine speed was only 120 revolutions per minute. This increase of 696 lbs. in the circulation of ammonia per 24 hours, together with the smaller consumption of steam (owing to the diminution in the speed of the engine) is due entirely to sufficient time being allowed the ammonia gas in its passage through the suction valves to maintain its suction pressure of 10 lbs., at which pressure 1 lb. of ammonia gas only occupies 11.794 cubic feet. If the piston traveled quicker than the above speed it sucked the gas instead of allowing it to follow by its own pressure, and thereby reduced the pressure to (in the cases of diagrams 11 and 12) 5 lbs., at which pressure 1 lb. of ammonia gas occupies 14.828 cubic feet, and the pumping capacity of the compressor, as far as the weight of ammonia circulated is concerned, is thereby reduced.

40 revolutions per minute $\times$ 4 = 160 cubic feet per minute = 230,400 cubic feet per 24 hours.

99 per cent. of this amount equals 228,096 cubic feet, and, as in the case of a thoroughly-

jacketed single-acting compressor, 21.4 per cent. instead of 30 per cent. has to be deducted. The effectual displacement in this case is 179,283 cubic feet per 24 hours.

We have already seen that 1 lb. of ammonia gas at 0° Fahr. and 10 lbs. (= 24¾ lbs. absolute pressure) = 11.794 cubic feet, and therefore the available 179,283 cubic feet = 15,201 lbs. of ammonia circulated per 24 hours.

The actual capacity of this vertical compressor is 230,400 cubic feet per 24 hours as against 224,640 in the case of the horizontal compressor when diagram 13 was taken, or an excess of only 5,760 cubic feet per 24 hours. Yet the increase in the amount of ammonia circulated amounted to (15,201 — 9,666) 5,535 lbs. of ammonia per 24 hours, which figures, if allowance is made for the 5,760 cubic feet excess capacity, are reduced to 5,042 lbs. This enormous increase of 5,042 lbs. in the weight of ammonia circulated is almost entirely due to the fact that the water-jacket on the compressor head of the vertical compressor causes a complete collapse of the clearance space gas, and thereby allows the suction-valves to open immediately the piston commences its return stroke.

Having ascertained the circulating capacity of our compressor we will now see what the freezing capacity of the plant is and how it could be improved.

We will suppose that the mean results of a 24-hour test were as follows:

Ammonia
- Gauge Pressure
  - Suction . . . . 10 lbs.
  - Discharge (Condensing) . . . . 140 lbs.
- Temperature
  - at Compressor
    - Suction . . . 8° Fahr.
    - Discharge . 251 Fahr.
  - at Discharge from Condenser, 62° Fahr.
  - at Refrig'ator Supply Manifold, 69° Fahr.
  - " " Discharge " 0° F.hr.

Brine
- Temperatures
  - Leaving Refrigerator . . . . . . 16½° Fahr.
  - Return to . . . . 31¾° Fahr.
- Revolutions of Pump per Minute 40
- Strength . . . . 22° Beaumé.

Revolutions of Compressor Engine per Minute 120

Diagram 13 represented the working of the compressor while the test was being made. The compressor piston displacement was 1.30 cubic feet per revolution.

The displacement of the brine pump piston was 0.8021 gallon per revolution.

Ammonia Figures.—Effectual Displacement.

Compressor: 120 revolutions per minute × 1.3 = 156 cubic feet per minute = 224,640 cubic feet per 24 hours. This amount less 27.5 per cent. = 162,864 cubic feet, and 30 per cent. deducted from that leaves 114,005 cubic feet effectual displacement per 24 hours.

Volume of Gas.

The gas as it entered the compressor was at a temperature of 8° Fahr. and under a gauge pressure of 10 lbs. (= 24.7 lbs. absolute pressure). By referring to Table VI. we see that 1 lb. of ammonia gas at 24¾ (24.75) lbs. absolute pressure and 8° Fahr. = 12.013 cubic feet and at 24.5 lbs. pressure and 8° Fahr. = 12.137 cubic feet. Our pressure was 24.7 lbs., or 0.05 lbs. less than 24¾, so, as there are 5, 5-100 difference between 24¾ and 24½, we divide the difference in the volume of the gas at those two pressures by

5 and add the quotient to the figures due to the pressure 24.75 lbs. Thus:

12.137 — 12.013 = 0.124; 0.124 ÷ 5 = 0.0248. 12.013 + 0.0248 = 12.0378 cubic feet = the volume of 1 lb. of ammonia gas at 8° Fahr. and 24.7 lbs. absolute pressure.

Ammonia Circulated per Twenty-four Hours.

The effectual displacement of the compressor was 162,864 cubic feet, and as the volume of one pound of the gas was 12.0378 cubic feet, the amount of ammonia circulated per 24 hours was (114,005 ÷ 12.0378) 9,470 lbs.

Refrigerating Efficiency.

We see by referring to Table III. (page 116) that the latent heat of ammonia at 9.86* lbs. gauge pressure is 561, therefore (9,470 × 561 =) 5,312,670 thermal units were absorbed by the ammonia in passing from the liquid to the gaseous state (*i. e.*, in ex-

---

* For all practical purposes these figures are near enough to 10 lbs.

panding), but the average results of the test show that the ammonia entered the refrigerator at a temperature of 69° Fahr. and that the gas left at a temperature of 0° Fahr.; it was therefore cooled down from 69° to 0°, or through 69 degrees, and as the specific heat of ammonia at suction pressures is 0.508, as already shown, it is evident (9,470 × 69 × .508) = 331,942 thermal units were thus utilized in cooling down the ammonia itself, and therefore, not being available for cooling down the brine, they must be deducted from the 5,312,670 thermal units credited to the ammonia, thus leaving (5,312,670 − 331,942 =) 4,980,728 effective thermal units, or (4,980,728 ÷ 284,800 =) 17.49 tons of ice per 24 hours.

Brine Figures.—Gallons Circulated.

The capacity of the brine pumps per revolution was 0.8021 gallon, and as it made 40 revolutions per minute, the volume of brine circulated was 0.8021 × 40 × 1440 = 46,200 gallons[*] per 24 hours.

---
[*] American gallons (= 8.34 lbs. of water).

POUNDS CIRCULATED.

The gravity of the brine was 22° Beaumé, and as brine at that strength weighs 9.84 lbs. per gallon, the number of pounds of brine circulated in the 24 hours was (46,200 × 9.84 =) 454,608.

DEGREES COOLED.

The average temperatures of the brine were:

Return — 31¾° Fahr.
Outgoing — 16½° Fahr.   Therefore the brine was cooled
15¼° Fahr.

TOTAL DEGREES EXTRACTED.

The total number of degrees Fahrenheit that were extracted from the brine were (454,608 × 15.25 =) 6,932,772.

CHAPTER X.

WE have shown previously that the specific heat of 22° Beaumé brine is 0.765, therefore the number of thermal units extracted were (6,932,772 × 0.705 =) 4,887,604, or (4,887,604 ÷ 284,800) 17.16 tons of ice per 24 hours. These figures give 0.33 ton of ice per 24 hours less than we obtained from the ammonia figures. This is a result that must always be looked for, as no insulation is perfectly non-conducting, and the air surrounding the refrigerator, etc., is always cooled more or less according to circumstances. The heat imparted to the refrigerator, etc., in this way is a varying amount and can not, under ordinary circumstances, be accurately estimated. It will have been noticed in the average ammonia temperatures that the liquid anhydrous ammonia was heated from 62° Fahr. up to 69° Fahr. in its passage from the condenser to the refrigerator supply manifold. We will now see what effect this rise in temperature had on the capacity of the plant.

Loss Due to Heating of Liquid Ammonia.

We have just figured that 5,312,670 thermal units were absorbed by the ammonia in passing from the liquid to the gaseous state, and that 331,942 thermal units of that amount had to be deducted for loss due to cooling the ammonia itself from 69° Fahr. to 0° Fahr.

Let it now be assumed that the temperature of the liquid ammonia remained at its condensing temperature of 62° Fahr. and our figures will be: 9,470 (lbs. of ammonia) × 62 × 0.508 = 298,267 thermal units required to cool the ammonia itself from 62° Fahr. to 0° Fahr., and therefore the number of thermal units available for cooling the brine would be (5,312,670 − 298,267 =) 5,014,403, or 17.61 tons of ice per 24 hours. These figures show that the seven degrees Fahrenheit that the ammonia was heated in its passage from the condenser to the refrigerator represented a loss in the refrigerating efficiency of the plant of (17.61 − 17.49 =) 0.12, or one-eighth of a ton of ice per 24 hours.

Loss Due to Heating of Ammonia Gas.

A glance at the average figures again will also show that the ammonia gas in its passage from the refrigerator to the compressor was heated eight degrees Fahrenheit—the gas entering the compressor at a temperature of $8°$ instead of $0°$. To determine what was the lost refrigerating effect in this case it will be necessary to calculate how many pounds of ammonia would have been circulated by the compressor had the temperature of the ammonia gas remained at $0°$ until it entered the compressor. Reference to Table V. (page 127) shows that 1 lb. of ammonia gas at 24.5 lbs. absolute pressure and $0°$ Fahr. has a volume of 11.917 cubic feet, and at 24.75 lbs. and $0°$ Fahr. 11.794 cubic feet;

therefore, at the absolute pressure of 24.7 lbs., the volume of 1 lb. of ammonia gas would be 11.8186 cubic feet. The effectual displacement of the compressor was 114,005 cubic feet per 24 hours, so the number of lbs. of ammonia circulated would be (114,005 ÷ 11.8186 =) 9,646 per 24 hours. The latent heat of vaporization we have already seen was 561, therefore (9,646 × 561 =) 5,411,406 thermal units would be absorbed by the ammonia. But the temperatures of the ammonia at the supply and discharge manifolds of the refrigerator were respectively 69° and 0° Fahr., and, consequently, as the ammonia itself had to be cooled sixty-nine degrees, the available number of thermal units would be reduced to (5,411,406 − (9,646 × 69 × 0.508) =) 5,073,244, or (5,073,244 ÷ 284,800 =) 17.81 tons of ice per 24 hours, showing that the loss due to the superheating of the gas only eight degrees in its passage from the refrigerator to the compressor amounted to (17.81 − 17.49 =) 0.32 ton, or about one-third of a ton of ice per 24 hours.

If the liquid anhydrous ammonia piping between the condenser and the refrigerator and the ammonia gas piping between the refrigerator and compressor had been covered with a thoroughly non-conducting material, the refrigerating efficiency of the plant would have been:

Gas entering Compressor at 0° Fahr............... 9,646 lbs.
5,411,406 Thermal units.
Ammonia cooled from 62° to 0° Fahr. (9,646 × 62 × 0.508).................... 303,810 " "

Effective Thermal Units = 5,107,596

or (5,107,596 ÷ 284,800 =) 17.93 tons of ice—being an increase of (17.93 — 17.49 =) 0.44, or nearly half a ton of ice per 24 hours.

As the question of condensing water has been fully discussed previously, it is considered unnecessary to go further into figures in relation to this part of the subject.

CHAPTER XI.

CALCULATION OF THE MAXIMUM CAPACITY OF A MACHINE.

As the capacity of a machine is proportional to the quantity of anhydrous ammonia circulated, it is evident that if the ammonia valves are regulated so as to give a brine temperature of 0° Fahr., the refrigerating efficiency expressed in tons of ice will not be nearly so great as when the valves are adjusted for a 28° Fahr. brine temperature. The amount of anhydrous ammonia circulated at the former temperature would only be one-half the weight circulated at the latter temperature.

If the brine temperature were above 28° Fahr. it would be incapable of doing practical refrigerating work—that is, the temperature would be too high to freeze water sufficiently quick to be of any practical value.

Twenty-eight degrees Fahrenheit is therefore the highest practical brine temperature, and in order to maintain that the ammonia must boil at 14° Fahr., which latter temperature is obtained by regulating the ammonia valves so that a suction-gauge pressure of 28¼ lbs. is maintained.

Therefore, in calculating the maximum capacity of a machine we must figure upon the suction-gauge pressure being 28¼ lbs. and the suction temperature, say, 20° Fahr. at the point where the gas enters the compressor.

Preparation of Anhydrous Ammonia.

The principal parts of the apparatus necessary for the production of anhydrous ammonia from 26° ammonia are:

(1) An iron cylinder (still) about 2 feet in diameter by 3 feet deep.

(2) An iron cylinder (column) about 10 inches in diameter by 2 feet high.

(3) A tank (condenser) about 3 feet in diameter by 4½ feet deep.

(4) Two iron cylinders (separators) about 10 inches in diameter by 5½ feet high.

(5) An iron vessel (dehydrator) about 3½ feet long by 2 feet broad and 2 feet deep.

Construction of Apparatus.

The apparatus should be of sufficient strength to withstand a pressure of 60 lbs. on the square inch. Its general arrangement is shown in section in Fig. 15, in which A is the still, the contents of which is heated by the steam coil, a. The ammonia gas, together with a little water vapor, pass off through b into the column B, and coming in contact with the plates c, the larger portion of the water separates and flows back into A by the pipe d, while the ammonia gas passes upwards through the holes e, and over to the condenser, C, After leaving the condenser the gas passes through the two separators D, D (where the water condensed in C separates) into the dryer, E, where, coming in contact with lime placed on the perforated plates f, it is rid of its last traces of moisture.

It is then drawn through the pipe *l* into the suction of the ammonia engine.

The plates in B are separated by, and rest on, the iron rings *i*. The head of the still and bottom end-plate of B, together with the connections *b* and *d*, may be conveniently cast in one piece.

CONDENSER-WORM.

An efficient worm for the condenser, C, may be cheaply and easily made of heavy lead pipe.

It is advisable to place a cock or valve on the connection between B and C, so that when the spent water is drawn from the still, the gas contained in the rest of the apparatus will not escape. However, it is not absolutely necessary to have a cock or valve at that point, because if the water is carefully run off no gas will escape.

After the still, A, has been charged it is slowly heated by the coil, *a*, to a temperature of about 212° Fahr. When the gauge, *k*, registers 25 to 30 lbs. pressure the valve connecting *l* with the suction of the compressor (of the ammonia engine) is opened and the engine run so as to maintain the pressure of 25 to 30 lbs.

Why Still is Worked under Pressure.

The reason for running the still under a pressure is to enable the contents of the still being heated up to, or slightly above, the normal boiling-point of water without allowing the water to boil—thus driving off the whole of the ammonia, while only a minimum quantity of the water is vaporized.

After the still has been heated for about an hour, a small quantity (about a teaspoonful) should be drawn off and tested with acid litmus paper, and as soon as it ceases to turn the paper blue it may be understood that the contents of the still have been exhausted of ammonia and that the charge is "spent."

Best Test for Ammonia.

A better method for telling when the charge is spent, is to have a small cock in the head of the still, and, opening it slightly, test the escaping vapors with a piece of turmeric paper. If the paper is turned brown, the whole of the ammonia has not been driven off, but if it still retains its yellow color the charge is thoroughly exhausted.

The spent water is run off from the still by the cock g, and after the still has cooled down it is ready for re-charging.

Water from Separators.

Very little water accumulates in the separators D, D, if the pressure in the still is carefully watched, but the cocks h, h should be cautiously opened (care being taken that no gas escapes) after about the fifth or sixth distillation, and if any water runs out it should be saved, as it will be saturated with ammonia gas, and therefore ought not to be thrown away, but should be placed in the drum containing the 26° ammonia.

Lime for Dehydrator.

The lime in E should be examined occasionally by removing the hand-hole plate, F, and if it has slaked to any great extent the cover on E should be removed and the plates *f* taken out and replenished with newly burnt lime broken in pieces about the size of a hen's egg. The lime should not be laid more than one layer deep on each plate.

The amount of 26° ammonia that has to be distilled in order to obtain a given quantity of anhydrous ammonia can be determined by the use of Table II.

Yield of Anhydrous from 26° Ammonia.

Let it be supposed that 50 gallons of anhydrous ammonia are required. By referring to the table it is seen, under the heading "Per Cent. by Volume," that 26° ammonia contains 38.5 per cent. of anhydrous ammonia, therefore, as 50 gallons of anhydrous ammonia are required it will be necessary to distill (38.5 : 50 :: 100) 130 gallons of 26° ammonia.

It is, of course, always advisable to try the strength of the 26° ammonia, as it is liable

TABLE II.

\multicolumn{3}{c	}{Solution.}	\multicolumn{5}{c}{Anhydrous Ammonia}
\multicolumn{2}{c|}{Weight of In.}						
Degrees Beaumé.	Pounds per Gallon.	Boiling Point.	Volume of Gas (at 32° Fahr., and Atmospheric Pressure) In one Volume of the Solution.	Pounds in One Gallon of the Solution.	Per Cent. by Volume.	Per Cent. by Weight.
34.7	7.09	26°	494	3.077	59.5	43.4
32.8	7.17	38°	456	2.841	54.9	39.6
31.0	7.25	50°	419	2.610	50.7	36.0
29.0	7.34	62°	382	2.379	46.0	32.5
27.2	7.42	74°	346	2.156	41.7	29.1
26.0	7.48	83°	320	1.993	38.5	26.6
25.6	7.50	86°	311	1.937	37.5	25.8
23.7	7.59	98°	277	1.726	33.4	22.8
22.2	7.67	110°	244	1.520	29.4	19.7

to vary somewhat; and should it be found stronger or weaker (*i. e.*, lighter or heavier in gravity) than the supposed strength, an allowance can be made, by means of Table II., when calculating the quantity necessary to be distilled to yield a given quantity of anhydrous ammonia.

The cost of preparing anhydrous ammonia from 26° ammonia is very small, and the difference in the price between the "home prepared" and the "commercial" anhydrous will very soon pay for the cost of the apparatus.

In most works were freezing plants are in use there are ample large-sized piping, small tanks or odd pieces of apparatus lying in disuse which could be easily fitted together on the principle of Fig. 15, and at a total cost of, say, $150.

The price of commercial anhydrous ammonia is 44.88c. per lb., and the price of commercial 26° ammonia is 6c. per lb.

Twenty-six degree ammonia contains 26.6 per cent. by weight of anhydrous ammonia, therefore 3.76 lbs. of 26° ammonia give 1 lb. of anhydrous at a cost (irrespective of labor) of 22.56c.

Ammonia Refrigeration. 115

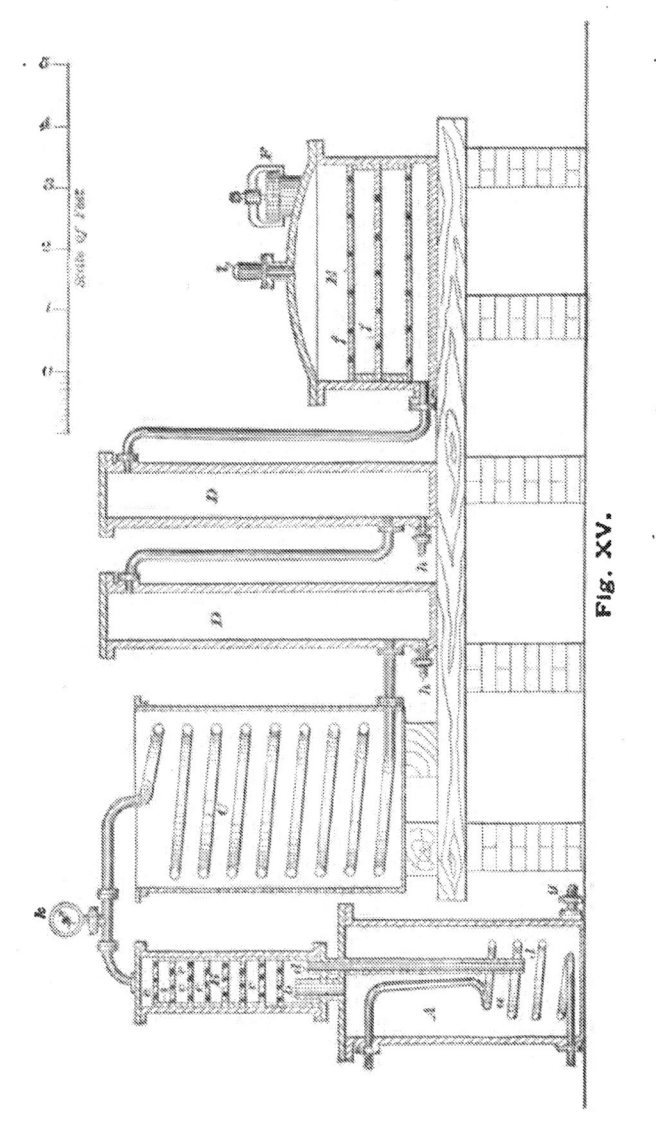

Fig. XV.

TABLE III.

Pressure Absolute	Pressure Gauge	Boiling Point °Fahr.	Latent Heat	Pressure Absolute	Pressure Gauge	Boiling Point	Latent Heat
10.69	−4.01	−40	579.7	58.00	43.30	28.9	537.6
11.00	−3.70	−39	579.1	59.41	44.71	30.0	536.9
12.31	−2.39	−35	576.7	60.00	45.30	30.6	536.5
13.00	−1.70	−32.7	575.3	61.50	46.80	32.0	535.7
14.13	−0.57	−30	573.7	62.00	47.30	32.3	535.5
14.70	∓0.00	−28.5	572.3	63.00	48.30	33.0	535.0
15.00	+0.30	−27.8	571.7	64.00	49.30	33.7	534.6
16.17	1.47	−25	570.7	65.93	51.23	35.0	533.8
16.71	2.01	−22	568.9	67.00	52.30	35.8	533.3
17.00	2.30	−21.8	568.7	69.00	54.30	37.2	532.4
18.45	3.75	−20	567.7	71.00	56.30	38.6	531.5
19.00	4.30	−18.9	567.0	73.00	58.30	40.0	530.6
20.99	6.29	−15	564.6	74.07	59.37	41.0	530.0
21.27	6.57	−13	563.4	75.00	60.30	41.5	529.7
22.10	7.40	−12	562.8	76.00	61.30	42.2	529.2
22.93	8.23	−11	562.2	78.00	63.30	43.4	528.5
23.77	9.07	−10	561.6	80.66	65.96	45.0	527.5
24.56	9.86	−9	561.0	88.96	74.26	50.0	524.3
25.32	10.62	−8	560.4	92.00	77.30	51.4	523.4
26.08	11.38	−7	559.8	95.00	80.30	53.2	522.3
26.84	12.14	−6	559.2	97.93	83.23	55.0	521.1
27.57	12.87	−5	558.5	100.00	85.30	56.1	520.4
28.09	13.39	−4	557.9	104.84	90.14	59.0	518.6
28.64	13.94	−3	557.3	107.60	92.90	60.0	517.9
29.17	14.47	−2	556.7	110.00	95.30	61.1	517.2
29.76	15.06	−1	556.1	115.00	100.30	63.5	515.7
30.37	15.67	∓0 (zero)	555.5	118.03	103.33	65.0	515.3
31.00	16.30	+1.4	554.6	119.70	105.00	66.0	514.1
32.00	17.30	3.5	553.4	123.59	108.89	68.0	512.8
33.66	18.96	5	552.4	125.20	112.50	69.0	512.2
35.00	20.30	5.9	551.9	127.21	114.51	70.0	511.5
36.00	21.30	7	551.2	138.70	124.00	74.5	508.6
37.00	22.30	8.2	550.5	141.25	127.55	75.0	508.3
38.55	23.85	10	549.3	144.67	129.97	77.0	507.0
39.00	24.30	10.6	549.0	149.70	135.00	78.5	506.0
40.00	25.30	12	548.1	154.11	139.41	80.0	504.7
42.20	27.50	14	546.8	161.70	147.00	82.5	503.5

TABLE III.—Continued.

Pressure (Absolute)	Pressure (Gauge)	Boiling Point °Fahr.	Latent Heat	Pressure (Absolute)	Pressure (Gauge)	Boiling Point	Latent Heat
42.93	28.23	15	546.3	165.70	151.00	84.5	502.1
44.00	29.30	16	545.6	166.70	152.00	84.9	501.8
45.00	30.30	17	545.0	167.86	153.16	85.4	501.6
46.00	31.30	18.1	544.3	168.30	153.60	85.7	501.2
47.00	32.30	19.1	543.7	168.70	154.00	86.0	500.8
47.95	33.25	20	543.1	175.70	161.00	88.5	499.5
49.00	34.30	21.1	542.5	182.80	168.10	90.0	498.1
50.00	35.30	22.3	541.7	194.80	180.10	95.0	495.3
50.67	35.97	23	541.3	204.70	190.00	98.0	493.3
51.00	36.30	23.3	541.1	215.14	200.44	100.0	491.5
52.00	37.30	24	540.7	224.40	209.70	104.0	489.4
53.43	38.73	25	540.0	257.20	242.50	113.0	483.4
54.00	39.30	25.5	539.7	293.20	278.50	122.0	476.4
55.00	40.30	26.3	539.3	333.10	318.40	131.0	471.4
56.00	41.30	27.1	538.7	377.20	352.50	140.0	465.4
57.00	42.30	28	538.2				

TABLE IV.

Temperature of Suction = 0° Fahr.

Absolute Condensing Pressure / Absolute Suction Pressure.

	20	22	25	27	30	32	35	37	40	24	45
90	199	184	165	153	138	129	116	109	98	92	83
95	208	193	173	161	146	137	124	118	105	99	90
100	216	201	181	169	153	144	131	123	113	105	97
105	224	208	188	177	161	151	137	130	119	113	103
110	232	215	196	183	166	155	145	137	126	119	109
115	239	223	203	191	174	164	151	143	132	125	115
120	245	230	211	197	181	171	158	149	138	131	121
125	253	237	216	204	187	177	164	156	144	137	127
130	261	244	222	210	193	183	169	161	150	142	132
135	266	250	229	216	199	189	175	167	155	148	138
140	273	256	235	222	205	194	181	172	161	155	143
145	279	262	240	228	210	197	186	178	166	158	147
150	285	268	246	233	216	206	191	183	171	164	153
155	291	273	252	239	221	211	197	188	176	169	158
160	296	279	257	244	226	216	202	193	181	173	163
165	302	285	262	249	232	221	206	198	185	178	167

Temperature of Suction = 5° Fahr.

Absolute Condensing Pressure / Absolute Suction Pressure.

	20	22	25	27	30	32	35	37	40	42	45
90	206	191	172	160	145	135	123	115	104	98	89
95	215	200	180	168	153	143	130	122	111	105	96
100	223	208	186	176	160	151	138	130	119	112	103
105	231	216	195	183	167	158	145	137	125	119	109
110	239	223	203	190	174	165	151	143	132	125	115
115	247	231	210	198	181	171	159	150	139	132	122
120	254	238	218	204	188	178	163	156	145	137	127
125	261	245	222	211	194	184	170	163	150	143	133
130	268	251	230	217	200	190	176	168	156	149	139
135	273	258	236	223	206	196	182	174	162	155	145
140	281	264	242	229	212	202	188	179	167	160	150
145	287	270	248	235	218	207	193	185	172	165	155
150	293	276	254	241	223	213	198	190	178	170	160
155	299	282	259	246	229	218	204	195	183	175	165
160	305	287	265	252	234	223	209	200	188	180	170
165	311	293	270	257	239	229	214	205	192	185	173

Ammonia Refrigeration.

TABLE IV.—Continued.

Temperature of Suction = 10° Fahr.

Absolute Suction Pressure.

Absolute Condensing Pressure.	20	22	25	27	30	32	35	37	40	42	45
90	213	198	178	167	151	141	129	121	110	104	96
95	222	207	187	175	159	150	136	129	118	111	102
100	231	215	195	183	167	157	144	136	125	118	109
105	239	223	202	190	174	164	151	143	132	125	115
110	247	229	210	197	181	172	158	150	139	132	122
115	254	238	217	205	188	178	164	156	145	138	128
120	261	245	226	211	195	185	171	163	151	144	134
125	269	252	231	218	201	191	177	169	157	150	140
130	275	259	237	224	207	197	183	175	163	155	145
135	282	266	244	231	214	203	189	181	168	161	151
140	289	272	250	237	219	209	195	186	174	167	156
145	295	278	256	244	225	211	200	192	179	172	162
150	301	284	262	248	231	220	205	197	185	177	167
155	307	290	266	254	236	225	211	202	190	182	172
160	313	295	273	259	241	231	216	207	195	187	176
165	319	301	278	265	247	236	221	212	199	192	181

Temperature of Suction = 15° Fahr.

Absolute Suction Pressure.

Absolute Condensing Pressure.	20	22	25	27	30	32	35	37	40	42	45
90	221	205	185	173	158	148	135	127	117	110	101
95	230	214	194	182	166	156	143	135	124	117	108
100	238	222	202	189	173	164	151	142	131	124	115
105	246	230	209	197	181	171	158	150	138	131	121
110	254	238	217	204	188	178	164	156	145	138	128
115	262	246	224	212	195	182	171	163	152	144	134
120	269	253	233	218	202	192	178	169	158	150	140
125	276	260	238	225	208	198	184	176	163	156	146
130	283	267	245	232	214	204	191	181	170	162	152
135	290	273	251	238	221	210	196	187	175	168	158
140	297	279	257	244	226	216	202	193	181	173	163
145	303	286	263	250	232	221	207	199	186	179	168
150	309	292	269	256	238	227	213	204	192	184	173
155	315	298	275	261	244	232	218	209	197	189	178
160	321	304	281	267	249	238	223	214	202	194	183
165	327	309	286	272	254	243	228	219	206	199	188

TABLE IV.—Continued.

Temperature of Suction = 20° Fahr.

Absolute Suction Pressure.

Absolute Condensing Pressure.	20	22	25	27	30	32	35	37	40	42	45
90	228	212	192	180	164	154	141	133	123	116	106
95	237	221	201	189	172	163	149	141	130	123	114
100	245	230	209	196	180	171	157	149	137	131	121
105	253	237	217	203	188	178	164	156	144	138	128
110	262	245	224	211	195	185	171	162	150	144	134
115	269	253	231	219	202	192	178	169	158	151	140
120	277	260	240	226	209	198	185	176	164	157	145
125	284	267	245	233	215	205	191	183	170	163	152
130	291	274	252	239	222	211	197	188	176	169	158
135	298	281	260	245	228	219	203	194	182	174	163
140	305	287	265	251	234	223	209	200	188	181	169
145	311	294	271	258	240	226	214	205	193	185	175
150	317	300	277	263	245	235	220	211	198	191	180
155	323	306	283	269	251	240	225	216	203	196	185
160	329	312	288	275	256	245	230	221	209	201	190
165	335	317	294	280	262	251	235	226	213	206	195

Temperature of Suction = 25° Fahr.

Absolute Suction Pressure.

Absolute Condensing Pressure.	20	22	25	27	30	32	35	37	40	42	45
90	235	219	199	186	171	161	148	140	129	122	111
95	244	228	207	195	179	169	155	148	136	129	120
100	252	237	216	203	187	177	163	155	144	137	127
105	261	245	224	211	194	183	171	162	150	144	134
110	269	251	230	218	200	191	178	169	155	150	140
115	277	260	239	226	209	198	184	176	164	157	147
120	284	268	247	232	216	205	191	182	171	163	153
125	292	275	253	240	222	212	197	189	177	169	159
130	299	282	259	246	229	218	204	195	183	175	165
135	306	289	267	253	235	224	210	201	188	181	171
140	313	295	271	259	241	230	216	207	194	187	176
145	319	301	278	265	247	236	221	212	200	192	181
150	325	308	284	271	253	242	227	218	205	197	187
155	332	314	290	277	258	247	232	223	210	203	192
160	338	320	296	282	264	253	237	228	216	208	197
	344	324	302	288	269	258	243	233	220	213	201

Ammonia Refrigeration.

TABLE IV.—Continued.

Temperature of Suction = 30° Fahr.

Absolute Suction Pressure.

Absolute Condensing Pressure.	20	22	25	27	30	32	35	37	40	42	45
90	242	226	206	193	177	167	154	146	134	128	118
95	251	235	214	202	185	176	162	154	142	136	125
100	260	244	223	210	193	184	170	161	150	143	133
105	269	252	231	218	201	191	177	169	157	150	140
110	277	260	239	226	208	199	184	176	164	157	147
115	285	268	246	233	216	205	191	183	171	164	153
120	292	275	255	240	223	212	198	189	177	170	159
125	300	283	260	247	230	219	204	196	183	176	165
130	307	290	267	254	236	225	211	202	190	182	171
135	314	297	274	260	242	232	217	208	195	188	177
140	321	303	280	266	248	237	223	214	201	193	183
145	327	309	286	273	254	240	228	219	207	199	188
150	334	316	292	278	260	249	234	225	212	204	193
155	340	322	298	284	266	255	240	230	217	210	199
160	346	328	304	290	271	260	245	234	223	215	203
165	352	334	310	296	277	266	250	241	230	220	208

Temperature of Suction = 32° Fahr.

Absolute Suction Pressure.

Absolute Condensing Pressure.	20	22	25	27	30	32	35	37	40	42	45
90	245	229	209	196	179	170	157	148	137	130	121
95	254	238	217	205	188	178	165	157	145	138	128
100	263	247	225	213	196	186	173	164	153	145	135
105	272	257	234	221	204	194	180	172	159	152	142
110	280	263	241	228	211	201	187	178	167	159	149
115	288	271	249	236	218	208	194	185	174	166	155
120	295	278	256	243	226	215	201	192	180	172	162
125	303	286	263	250	232	222	207	199	186	178	168
130	310	293	270	256	239	228	213	204	192	184	174
135	317	300	277	263	245	234	220	211	198	190	180
140	324	306	283	269	251	240	226	216	204	196	185
145	330	313	289	276	257	243	231	222	209	203	191
150	337	319	295	281	263	252	237	227	215	207	196
155	343	325	301	287	269	257	243	233	220	212	201
160	350	331	307	293	274	263	248	238	226	218	206
165	355	337	313	299	280	268	253	244	233	223	211

TABLE IV.—*Continued.*

Temperature of Suction = 35° Fahr.

Absolute Condensing Pressure.	\multicolumn{11}{c}{Absolute Suction Pressure.}
	20	23	25	27	30	33	35	37	40	42	45
90	249	233	213	200	182	174	160	152	141	134	124
95	259	243	221	209	192	182	168	160	148	142	132
100	268	251	229	217	200	190	176	168	156	149	139
105	276	259	238	225	208	198	184	175	163	156	146
110	285	267	246	233	215	205	191	182	170	163	153
115	292	275	253	240	223	212	198	189	178	170	159
120	300	285	260	247	230	219	205	196	184	176	166
125	308	290	268	254	237	226	211	203	190	182	172
130	315	297	274	261	243	232	217	208	196	188	178
135	322	304	281	268	249	239	222	215	202	194	184
140	329	311	288	274	255	244	230	221	208	200	189
145	335	317	294	280	262	247	235	226	213	205	195
150	341	324	300	286	268	257	241	232	219	211	200
155	348	330	306	292	273	262	247	237	224	217	205
160	354	336	312	298	279	268	252	243	230	222	210
165	360	342	318	303	284	273	257	248	235	227	215

TABLE V.

Pounds per Square Inch Absolute Pressure.

Temp. °Fahr.	14.7	Temp. °Fahr.	14.7	Temp. °Fahr.	14.7	Temp. °Fahr.	14.7	Temp. °Fahr.	14.7
\multicolumn{10}{c}{Volume in Cubic Feet of One Pound Weight of Gas.}
0	20.001	11	20.505	21	20.954	31	21.403
1	20.042	12	20.545	22	20.994	32	21.457
2	20.096	13	20.589	23	21.049	33	21.498
3	20.137	14	20.641	24	21.089	34	21.539
4	20.178	15	20.680	25	21.130	35	21.593
5	20.233	16	20.722	26	21.183	36	21.634
6	20.273	17	20.777	27	21.226	37	21.675
7	20.314	18	20.818	28	21.266	38	21.729
8	20.368	19	20.858	29	21.321	39	21.770
9	20.409	20	20.913	30	21.362	40	21.809
10	20.450								

Ammonia Refrigeration.

TABLE V.—Continued.

Temperature °Fahr.	\multicolumn{8}{c}{Pounds per Square Inch Absolute Pressure.}
	15	15¼	15½	15¾	16	16¼	16½	16¾
	\multicolumn{8}{c}{Volume in Cubic Feet of One Pound Weight of Gas.}
0	19.600	19.271	18.956	18.651	18.357	18.071	17.793	17.524
1	19.637	19.310	18.995	18.686	18.394	18.108	17.820	17.554
2	19.690	19.362	19.046	18.737	18.444	18.157	17.878	17.608
3	19.730	19.402	19.085	18.775	18.482	18.194	17.914	17.644
4	19.770	19.441	19.124	18.813	18.519	18.238	17.951	17.679
5	19.823	19.494	19.175	18.864	18.569	18.280	17.999	17.727
6	19.863	19.533	19.214	18.902	18.607	18.317	18.036	17.763
7	19.900	19.572	19.253	18.940	18.644	18.354	18.072	17.799
8	19.957	19.623	19.311	18.991	18.694	18.403	18.121	17.847
9	19.995	19.662	19.343	19.032	18.732	18.440	18.157	17.882
10	20.036	19.703	19.382	19.070	18.769	18.477	18.193	17.918
11	20.090	19.755	19.432	19.121	18.819	18.526	18.242	17.965
12	20.133	19.795	19.472	19.159	18.856	18.563	18.278	18.002
13	20.170	19.835	19.511	19.197	18.894	18.600	18.315	13.038
14	20.223	19.885	19.563	19.248	18.944	18.649	18.363	18.085
15	20.263	19.924	19.601	19.286	18.982	18.686	18.399	18.121
16	20.303	19.964	19.640	19.324	19.019	18.723	18.436	18.157
17	20.357	20.018	19.691	19.375	19.069	18.772	18.484	18.205
18	20.396	20.058	19.730	19.413	19.107	18.809	18.521	18.241
19	20.437	20.097	19.769	19.451	19.144	18.846	18.557	18.276
20	20.490	29.149	19.821	19.502	19.194	18.895	18.605	18.324
21	20.523	20.189	19.859	19.540	19.247	18.932	18.642	18.360
22	20.570	20.221	19.898	19.578	19.298	18.969	18.678	18.396
23	20.623	20.281	19.950	19.629	19.330	19.018	18.727	18.444
24	20.663	20.320	19.988	19.667	19.376	19.055	18.763	18.479
25	20.703	20.359	20.027	19.705	19.410	19.092	18.799	18.515
26	20.756	20.412	20.079	19.756	19.444	19.141	18.848	18.563
27	20.797	20.451	20.117	19.794	19.482	19.178	18.885	18.599
28	20.837	20.490	20.156	19.832	19.519	19.215	18.921	18.634
29	20.890	20.543	20.208	19.883	19.569	19.265	18.969	18.682
30	20.930	20.582	20.246	19.921	19.607	19.301	19.005	18.718
31	20.970	20.622	20.285	19.959	19.644	19.338	19.042	18.754
32	21.023	20.674	20.337	20.010	19.694	19.388	19.090	18.802
33	21.063	20.713	20.375	20.048	19.732	19.425	19.127	18.837
34	21.103	20.753	20.414	20.086	19.769	19.462	19.163	18.873
35	21.156	20.804	20.466	20.137	19.819	19.511	19.211	18.921
36	21.197	20.844	20.505	20.175	19.851	19.548	19.248	18.957
37	21.236	20.884	20.543	20.213	19.894	19.585	19.284	18.993
38	21.290	20.936	20.595	20.264	19.944	19.634	19.333	19.041
39	21.330	20.976	20.633	20.302	19.982	19.671	19.369	19.076
40	21.370	21.015	20.672	20.340	20.019	19.708	19.405	19.112

TABLE V.—*Continued.*

Temperature °Fahr.	\multicolumn{8}{c}{Pounds per Square Inch Absolute Pressure.}
	17	17¼	17½	17¾	18	18¼	18½	18¾
	\multicolumn{8}{c}{Volume in Cubic Feet of One Pound Weight of Gas.}
0	17.263	17.009	16.763	16.524	16.292	16.065	15.845	15.631
1	17.298	17.044	16.792	16.558	16.325	16.098	15.878	15.663
2	17.345	17.090	16.843	16.603	16.369	16.142	15.921	15.705
3	17.381	17.125	16.878	16.637	16.403	16.174	15.953	15.738
4	17.416	17.160	16.912	16.670	16.436	16.208	15.984	15.769
5	17.463	17.206	16.958	16.715	16.481	16.251	16.029	15.812
6	17.498	17.241	16.992	16.749	16.514	16.284	16.062	15.844
7	17.534	17.276	17.026	16.783	16.547	16.317	16.094	15.876
8	17.575	17.322	17.072	16.828	16.592	16.361	16.137	15.919
9	17.616	17.357	17.106	16.862	16.625	16.394	16.170	15.951
10	17.651	17.392	17.141	16.896	16.658	16.427	16.202	15.983
11	17.698	17.438	17.186	16.941	16.703	16.471	16.245	16.025
12	17.733	17.473	17.221	16.975	16.736	16.503	16.278	16.058
13	17.769	17.508	17.255	17.008	16.769	16.536	16.310	16.089
14	17.816	17.554	17.301	17.053	16.814	16.580	16.353	16.132
15	17.865	17.589	17.335	17.087	16.847	16.613	16.386	16.164
16	17.886	17.624	17.369	17.121	16.880	16.646	16.418	16.196
17	17.933	17.670	17.415	17.166	16.925	16.690	16.462	16.239
18	17.969	17.705	17.449	17.200	16.958	16.723	16.494	16.271
19	18.004	17.740	17.483	17.234	16.992	16.756	16.526	16.303
20	18.051	17.786	17.529	17.279	17.036	16.799	16.570	16.345
21	18.086	17.821	17.563	17.312	17.069	16.832	16.602	16.377
22	18.122	17.859	17.598	17.346	17.103	16.865	16.634	16.410
23	18.169	17.902	17.643	17.391	17.147	16.909	16.678	16.452
24	18.204	17.937	17.678	17.425	17.180	16.942	16.710	16.484
25	18.239	17.972	17.711	17.459	17.215	16.973	16.743	16.516
26	18.286	18.018	17.757	17.504	17.258	17.018	16.786	16.539
27	18.322	18.053	17.792	17.538	17.292	17.051	16.818	16.591
28	18.357	18.088	17.826	17.571	17.325	17.084	16.851	16.623
29	18.404	18.134	17.872	17.617	17.369	17.123	16.894	16.666
30	18.439	18.169	17.906	17.651	17.403	17.161	16.926	16.697
31	18.475	18.204	17.941	17.685	17.436	17.194	16.959	16.730
32	18.522	18.250	17.986	17.730	17.469	17.238	17.002	16.772
33	18.557	18.285	18.021	17.763	17.514	17.271	17.034	16.804
34	18.592	18.319	18.055	17.797	17.547	17.304	17.067	16.836
35	18.639	18.366	18.101	17.822	17.591	17.347	17.110	16.879
36	18.675	18.401	18.135	17.876	17.625	17.380	17.143	16.911
37	18.710	18.435	18.169	17.910	17.658	17.413	17.175	16.943
38	18.757	18.482	18.215	17.955	17.703	17.457	17.218	16.985
39	18.792	18.517	18.263	17.989	17.736	17.490	17.251	17.023
40	18.828	18.551	18.283	18.002	17.769	17.523	17.283	17.055

Ammonia Refrigeration.

TABLE V.—Continued.

Temperature °Fahr.	\multicolumn{8}{c}{Pounds per Square Inch Absolute Pressure.}
	19	19¼	19½	19¾	20	20¼	20½	20¾
	\multicolumn{8}{c}{Volume in Cubic Feet of One Pound Weight of Gas.}
0	15.421	15.220	15.022	14.828	14.641	14.451	14.279	14.104
1	15.454	15.251	15.052	14.859	14.671	14.487	14.308	14.132
2	15.496	15.292	15.093	14.899	14.711	14.526	14.347	14.172
3	15.528	15.324	15.124	14.930	14.741	14.551	14.376	14.200
4	15.559	15.355	15.157	14.960	14.771	14.584	14.405	14.229
5	15.602	15.396	15.196	15.001	14.811	14.625	14.444	14.268
6	15.633	15.427	15.227	15.031	14.841	14.655	14.474	14.297
7	15.665	15.459	15.257	15.061	14.871	14.684	14.503	14.326
8	15.707	15.500	15.298	15.102	14.911	14.724	14.542	14.364
9	15.738	15.530	15.329	15.132	14.941	14.754	14.571	14.393
10	15.770	15.563	16.360	15.163	14.971	14.783	14.600	14.419
11	15.812	15.604	15.401	15.203	15.011	14.823	14.639	14.461
12	15.844	15.635	15.432	15.238	15.041	14.852	14.669	14.490
13	15.875	15.666	15.462	15.264	15.071	14.882	14.698	14.519
14	15.917	15.708	15.504	15.304	15.111	17.920	14.737	14.557
15	15.949	15.739	15.534	15.340	15.141	17.950	14.771	14.586
16	15.980	15.770	15.565	15.365	15.171	14.981	14.796	14.615
17	16.021	15.812	15.606	15.406	15.211	15.020	14.835	14.652
18	16.054	15.843	15.637	15.436	15.241	15.050	14.864	14.682
19	16.086	15.874	15.668	15.466	15.271	15.080	14.893	14.711
20	16.128	15.916	15.709	15.507	15.311	15.119	14.932	14.741
21	16.159	15.947	15.739	15.537	15.341	15.149	14.961	14.779
22	16.191	15.978	15.770	15.568	15.371	15.178	14.991	14.808
23	16.244	16.020	15.811	15.608	15.411	15.218	15.030	14.846
24	16.265	16.051	15.842	15.638	15.441	15.252	15.058	14.875
25	16.296	16.082	15.873	15.669	15.471	15.277	15.088	14.904
26	16.338	16.124	15.913	15.709	15.511	15.317	15.127	14.943
27	16.370	16.155	15.945	15.740	15.541	15.346	15.152	14.972
28	16.401	16.186	15.975	15.770	15.571	15.376	15.186	15.000
29	16.444	16.227	16.002	15.811	15.611	15.415	15.226	15.039
30	16.475	16.259	16.047	15.841	15.642	15.444	15.254	15.068
31	16.502	16.290	16.078	15.871	15.671	15.475	15.283	15.097
32	16.549	16.331	16.119	15.912	15.711	15.514	15.322	15.135
33	16.580	16.363	16.150	15.942	15.742	15.544	15.352	15.104
34	16.612	16.394	16.181	15.973	15.771	15.573	15.381	15.193
35	16.654	16.435	16.222	16.013	15.811	15.613	15.420	15.231
36	16.686	16.466	16.253	16.044	15.841	15.642	15.449	15.261
37	16.717	16.497	16.283	16.074	15.871	15.672	15.479	15.294
38	16.759	16.539	19.324	16.109	15.911	15.712	15.518	15.323
39	16.791	16.570	16.355	16.145	15.942	15.741	15.552	15.354
40	16.824	16.602	16.386	16.175	15.971	15.771	15.576	15.386

TABLE V.—Continued.

Temperature °Fahr.	21	21¼	21½	21¾	22	22¼	22½	22¾
\multicolumn{9}{c}{Pounds per Square Inch Absolute Pressure.}
\multicolumn{9}{c}{Volume in Cubic Feet of One Pound Weight of Gas.}
0	13.934	13.768	13.605	13.446	13.292	13.140	12.992	12.847
1	13.963	13.796	13.633	13.474	13.319	13.167	13.019	12.873
2	14.001	13.833	18.670	13.511	13.356	13.203	13.054	12.905
3	14.029	13.863	13.698	13.538	13.383	13.230	13.081	12.934
4	14.058	13.890	13.726	13.566	13.410	13.257	13.108	12.961
5	14.096	13.928	13.763	13.603	13.447	13.293	13.143	12.996
6	14.125	13.956	13.791	13.630	13.474	13.320	13.170	13.023
7	14.153	13.984	13.819	13.658	13.501	13.347	13.196	13.049
8	14.191	14.022	13.856	13.695	13.538	13.383	13.232	13.084
9	14.220	14.050	13.884	13.722	13.565	13.410	13.259	13.111
10	14.249	14.078	13.912	13.750	13.594	13.437	13.285	13.137
11	14.287	14.116	13.949	13.787	13.629	13.473	13.321	13.172
12	14.315	14.144	13.977	13.814	13.656	13.500	13.348	13.199
13	14.344	14.172	14.005	13.842	13.683	13.527	13.374	13.225
14	14.382	14.210	14.042	13.879	13.719	13.563	13.410	13.260
15	14.410	14.237	14.070	13.906	13.747	13.590	13.436	13.287
16	14.439	14.266	14.098	13.934	13.774	13.617	13.463	13.313
17	14.477	14.304	14.135	13.971	13.801	13.653	13.499	13.348
18	14.506	14.332	14.163	13.998	13.838	13.680	13.525	13.374
19	14.534	14.360	14.191	14.026	13.865	13.706	13.552	13.401
20	14.572	17.398	14.228	14.063	13.901	13.742	13.588	13.436
21	14.601	14.426	14.256	14.090	13.929	13.769	13.614	13.462
22	14.629	14.455	14.284	14.118	13.956	13.796	13.641	13.489
23	14.668	14.492	14.321	14.154	13.992	13.832	13.676	13.524
24	14.696	14.520	14.349	14.182	14.020	13.859	13.703	13.550
25	14.725	14.549	14.377	14.210	14.047	13.886	13.730	13.577
26	14.763	14.586	14.414	14.246	14.083	13.922	13.765	13.612
27	14.787	14.615	14.442	14.274	14.110	13.949	13.792	13.638
28	14.825	14.643	14.470	14.302	14.138	13.976	13.819	13.665
29	14.858	14.680	14.507	14.338	14.174	14.012	13.854	13.700
30	14.887	14.709	14.535	14.366	14.201	14.039	13.881	13.726
31	14.915	14.737	14.563	14.389	14.229	14.066	13.908	13.752
32	14.953	14.775	14.600	14.430	14.265	14.102	13.943	13.788
33	14.982	14.803	14.628	14.458	14.292	14.129	13.970	13.814
34	15.010	14.831	14.656	14.485	14.319	14.156	13.996	13.840
35	15.049	14.869	14.693	14.522	14.356	14.192	14.032	13.876
36	15.077	14.897	14.721	14.548	14.383	14.219	14.059	13.902
37	15.106	14.925	14.749	14.577	14.410	14.246	14.085	13.928
38	15.144	14.963	14.786	14.614	14.447	14.282	14.121	13.963
39	15.172	14.991	14.814	14.642	14.474	14.309	14.148	13.990
40	15.201	15.015	14.842	14.669	14.501	14.336	14.174	14.016

Ammonia Refrigeration.

TABLE V.—Continued.

Temperature °Fahr.	\multicolumn{8}{c}{Pounds per Square Inch Absolute Pressure.}
	23	23¼	23½	23¾	24	24¼	24½	24¾
	\multicolumn{8}{c}{Volume in Cubic Feet of One Pound Weight of Gas.}
0	12.706	12.567	12.421	12.299	12.169	12.041	11.917	11.794
1	12.732	12.588	12.457	12.324	12.194	12.066	11.941	11.819
2	12.767	12.627	12.491	12.357	12.227	12.098	11.974	11.851
3	12.793	12.653	12.516	12.383	12.252	12.123	11.998	11.875
4	12.819	12.674	12.542	12.412	12.277	12.148	12.023	11.899
5	12.854	12.713	12.576	12.442	12.310	12.181	12.055	11.932
6	12.880	12.739	12.601	12.467	12.335	12.206	12.080	11.956
7	12.906	12.765	12.627	12.492	12.360	12.230	12.104	11.981
8	12.941	12.799	12.661	12.526	12.394	12.263	12.137	12.013
9	12.967	12.825	12.686	12.551	12.419	12.288	12.162	12.037
10	12.993	12.849	12.712	12.576	12.444	12.313	12.186	12.061
11	13.028	12.885	12.746	12.610	12.477	12.343	12.219	12.094
12	13.054	12.911	12.771	12.635	12.502	12.371	12.243	12.118
13	13.080	12.932	12.797	12.661	12.527	12.395	12.268	12.142
14	13.115	12.971	12.831	12.694	12.560	12.428	12.300	12.174
15	13.141	12.997	12.857	12.720	12.585	12.453	12.325	12.199
16	13.167	13.023	12.882	12.745	12.610	12.478	12.349	12.223
17	13.201	13.057	12.916	12.778	12.644	12.511	12.382	12.255
18	13.228	13.083	12.942	12.804	12.669	12.540	12.406	12.279
19	13.254	13.109	12.967	12.829	12.694	12.560	12.431	12.304
20	13.288	13.143	13.001	12.863	12.727	12.593	12.464	12.336
21	13.315	13.169	13.027	12.888	12.752	12.618	12.488	12.360
22	13.341	13.195	13.052	12.913	12.777	12.643	12.512	12.385
23	13.376	13.229	13.086	12.947	12.810	12.676	12.545	12.417
24	13.402	13.255	13.112	12.972	12.835	12.701	12.570	12.441
25	13.428	13.281	13.137	12.997	12.861	12.725	12.594	12.465
26	13.462	13.315	13.171	13.031	12.893	12.758	12.627	12.498
27	13.488	13.341	13.197	13.056	12.919	12.783	12.651	12.522
28	13.515	13.367	13.223	13.082	12.944	12.808	12.675	12.546
29	13.549	13.401	13.257	13.117	12.977	12.841	12.709	12.579
30	13.576	13.427	13.286	13.141	13.004	12.868	12.733	12.603
31	13.602	13.453	13.308	13.166	13.027	12.890	12.758	12.627
32	13.637	13.487	13.342	13.200	13.060	12.923	12.790	12.659
33	13.663	13.513	13.367	13.225	13.085	12.948	12.815	12.684
34	13.689	13.539	13.393	13.250	13.110	12.974	12.839	12.708
35	13.723	13.573	13.427	13.284	13.144	13.006	12.872	12.740
36	13.749	13.599	13.452	13.309	13.167	13.030	12.896	12.764
37	13.776	13.629	13.478	13.334	13.194	13.055	12.921	12.789
38	13.802	13.659	13.512	13.368	13.227	13.088	12.953	12.821
39	13.837	13.685	13.537	13.393	13.252	13.113	12.978	12.845
40	13.863	13.711	13.563	13.419	13.277	13.138	13.002	12.869

TABLE V.—*Continued.*

Temperature °Fahr.	\multicolumn{8}{c}{Pounds per Square Inch Absolute Pressure.}
	25	25¼	25½	25¾	26	26¼	26½	25¾
	\multicolumn{8}{c}{Volume in Cubic Feet of One Pound Weight of Gas.}
0	11.675	11.558	11.440	11.330	11.219	11.111	11.005	10.900
1	11.699	11.581	41.466	11.353	11.242	11.134	11.027	10.923
2	11.731	11.613	11.498	11.384	11.273	11.164	11.057	10.952
3	11.755	11.637	11.521	11.408	11.296	11.187	11.080	10.975
4	11.779	11.661	11.545	11.431	11.319	11.210	11.103	10.997
5	11.811	11.692	11.576	11.462	11.350	11.240	11.133	11.027
6	11.835	11.716	11.599	11.486	11.373	11.252	11.155	11.050
7	11.859	11.740	11.623	11.509	11.396	11.286	11.178	11.072
8	11.891	11.771	11.655	11.540	11.427	11.317	11.208	11.102
9	11.915	11.795	11.678	11.563	11.450	11.339	11.231	11.124
10	11.939	11.819	11.702	11.586	11.473	11.362	11.254	11.147
11	11.971	11.851	11.733	11.617	11.504	11.393	11.284	11.177
12	11.995	11.874	11.753	11.641	11.527	11.415	11.306	11.199
13	12.019	11.898	11.780	11.664	11.550	11.438	11.329	11.222
14	12.038	11.930	11.811	11.695	11.581	11.469	11.359	11.252
15	12.075	11.954	11.835	11.718	11.604	11.492	11.382	11.274
16	12.099	11.977	11.858	11.742	11.627	11.515	11.405	11.296
17	12.131	12.009	11.890	11.773	11.658	11.545	11.435	11.326
18	12.155	12.033	11.913	11.796	11.681	11.568	11.457	11.349
19	12.179	12.057	11.937	11.819	11.704	11.591	11.480	11.371
20	12.211	12.088	11.964	11.850	11.735	11.621	11.510	11.401
21	12.235	12.112	11.992	11.874	11.758	11.644	11.533	11.423
22	12.259	12.136	12.015	11.897	11.781	11.667	11.555	11.446
23	12.291	12.168	12.047	11.928	11.811	11.697	11.586	11.476
24	12.315	12.192	12.070	11.951	11.835	11.720	11.608	11.498
25	12.339	12.215	12.094	11.975	11.857	11.743	11.631	11.521
26	12.371	12.247	12.125	12.006	11.888	11.774	11.661	11.551
27	12.395	12.270	12.149	12.029	11.911	11.797	11.684	11.573
28	12.419	12.294	12.172	12.052	11.935	11.819	11.706	11.595
29	12.451	12.326	12.204	12.083	11.965	11.850	11.737	11.625
30	12.473	12.350	12.227	12.107	11.988	11.873	11.755	11.648
31	12.499	12.373	12.251	12.130	12.011	11.895	11.782	11.670
32	12.531	12.405	12.282	12.161	12.042	11.926	11.812	11.700
33	12.555	12.429	12.305	12.184	12.065	11.945	11.834	11.723
34	12.579	12.453	12.329	12.208	12.088	11.972	11.857	11.745
35	12.611	12.484	12.360	12.239	12.119	12.002	11.888	11.775
36	12.636	12.508	12.384	12.262	12.142	12.025	11.910	11.797
37	12.659	12.530	12.407	12.285	12.165	12.048	11.933	11.820
38	12.691	12.564	12.439	12.316	12.196	12.078	11.963	11.850
39	12.716	12.587	12.462	12.340	12.219	12.101	11.986	11.872
40	12.739	12.611	12.486	12.363	12.242	12.124	12.008	11.894

TABLE V.—Continued.

Temperature °Fahr.	\multicolumn{8}{c}{Pounds per Square Inch Absolute Pressure.}
	27	27¼	27½	27¾	28	28¼	28½	28¾
	\multicolumn{8}{c}{Volume in Cubic Feet of One Pound Weight of Gas.}
0	10.797	10.697	10.598	10.501	10.406	10.313	10.221	10.130
1	10.819	10.719	10.620	10.523	10.428	10.334	10.242	10.151
2	10.849	10.748	10.650	10.552	10.456	10.362	10.270	10.179
3	10.872	10.770	10.671	10.573	10.477	10.383	10.291	10.200
4	10.894	10.792	10.693	10.595	10.499	10.405	10.312	10.221
5	10.923	10.822	10.722	10.624	10.527	10.433	10.340	10.249
6	10.946	10.844	10.744	10.645	10.549	10.454	10.361	10.270
7	10.964	10.866	10.766	10.667	10.570	10.475	10.382	10.290
8	10.997	10.895	10.795	10.696	10.599	10.504	10.410	10.318
9	11.019	10.917	10.816	10.717	10.620	10.525	10.431	10.340
10	11.042	10.939	10.838	10.739	10.642	10.546	10.452	10.360
11	11.071	10.968	10.868	10.768	10.670	10.578	10.480	10.388
12	11.094	10.990	10.889	10.789	10.692	10.596	10.501	10.409
13	11.116	11.012	10.911	10.811	10.713	10.617	10.522	10.430
14	11.145	11.042	10.940	10.840	10.742	10.645	10.551	10.457
15	11.167	11.064	10.962	10.862	10.763	10.666	10.572	10.478
16	11.190	11.086	10.984	10.883	10.785	10.688	10.593	10.499
17	11.219	11.115	11.013	10.912	10.813	10.716	10.621	10.527
18	11.242	11.137	11.035	10.934	10.835	10.737	10.642	10.548
19	11.264	11.160	11.056	10.955	10.859	10.759	10.663	10.569
20	11.294	11.189	11.085	10.984	10.885	10.787	10.691	10.596
21	11.316	11.211	11.107	11.006	10.906	10.808	10.712	10.617
22	11.338	11.233	11.129	11.027	10.927	10.830	10.733	10.638
23	11.367	11.262	11.158	11.056	10.956	10.858	10.761	10.666
24	11.390	11.284	11.180	11.078	10.977	10.879	10.782	10.687
25	11.412	11.306	11.202	11.099	10.999	10.900	10.803	10.708
26	11.442	11.335	11.231	11.128	11.027	10.928	10.831	10.736
27	11.464	11.356	11.253	11.150	11.049	10.950	10.852	10.756
28	11.486	11.379	11.275	11.171	11.070	10.971	10.873	10.777
29	11.516	11.405	11.308	11.200	11.099	10.999	10.901	10.805
30	11.538	11.431	11.325	11.222	11.120	11.021	10.922	10.826
31	11.560	11.453	11.347	11.244	11.142	11.042	10.944	10.847
32	11.590	11.482	11.376	11.272	11.170	11.070	10.972	10.875
33	11.612	11.504	11.398	11.294	11.192	11.091	10.993	10.896
34	11.634	11.526	11.420	11.316	11.213	11.113	11.013	10.916
35	11.664	11.556	11.449	11.345	11.242	11.141	11.042	10.944
36	11.686	11.577	11.461	11.366	11.263	11.162	11.063	10.965
37	11.709	11.600	11.493	11.388	11.285	11.183	11.084	10.986
38	11.738	11.623	11.522	11.417	11.313	11.212	11.112	11.014
39	11.760	11.651	11.544	11.438	11.335	11.233	11.133	11.035
40	11.783	11.673	11.565	11.460	11.363	11.254	11.154	11.056

TABLE V.—*Continued.*

POUNDS PER SQUARE INCH ABSOLUTE PRESSURE.

Temp. °Fahr.	29	29¼	29½	29¾	Temp. °Fahr.	29	29¼	29½	29¾
\multicolumn{10}{c}{Volume in Cubic Feet of One Pound Weight of Gas.}

Temp.	29	29¼	29½	29¾	Temp.	29	29¼	29½	29¾
0	10.042	9.955	9.869	9.785	21	10.524	10.433	10.343	10.255
1	10.062	9.975	9.889	9.805	22	10.545	10.454	10.364	10.275
2	10.090	10.003	9.915	9.832	23	10.573	10.481	10.391	10.302
3	10.111	10.023	9.936	9.852	24	10.593	10.502	10.411	16.322
4	10.131	10.044	9.957	9.872	25	10.611	10.522	10.432	10.343
5	10.159	10.071	9.984	9.899	26	10.642	10.549	10.459	10.369
6	10.179	10.091	10.004	9.919	27	10.662	10.570	10.479	10.390
7	10.200	10.112	10.025	9.939	28	10.683	10.590	10.499	10.410
8	10.228	10.139	10.052	9.966	29	10.711	10.618	10.526	10.437
9	10.249	10.160	10.072	9.986	30	10.731	10.638	10.547	10.457
10	10.269	10.180	10.093	10.006	31	10.752	10.659	10.567	10.477
11	10.297	10.208	10.120	10.033	32	10.779	10.686	10.594	10.504
12	10.318	10.228	10.140	10.053	33	10.800	10.707	10.615	10.524
13	10.338	10.249	10.160	10.074	34	10.821	10.727	10.635	10.544
14	10.365	10.276	10.188	10.101	35	10.848	10.755	10.662	10.571
15	10.385	10.296	10.208	10.121	36	10.869	10.775	10.682	10.591
16	10.407	10.317	10.228	10.141	37	10.890	10.796	10.703	10.611
17	10.435	10.344	10.255	10.168	38	10.918	10.823	10.730	10.638
18	10.455	10.365	10.276	10.188	39	10.938	16.843	10.750	10.658
19	10.476	10.385	10.296	10.208	40	10.959	10.864	10.770	10.679
20	10.504	10.413	10.323	10.235					

———*o*———

TABLE VI.

Temp. °Fahr.	30	30¼	30½	30¾	31	31¼	31½	31¾
	\multicolumn{8}{c}{Volume in Cubic Feet of One Pound Weight of Gas.}
0	9.701	9.620	9.540	9.461	9.374	9.307	9.232	9.159
1	9.722	9.640	9.560	9.481	9.403	9.327	9.251	9.178
2	9.748	9.666	9.586	9.507	9.429	9.352	9.277	9.203
3	9.768	9.686	9.606	9.527	9.448	9.371	9.296	9.222
4	9.788	9.706	9.625	9.546	9.468	9.391	9.315	9.241
5	9.813	9.733	9.651	9.572	9.493	9.416	9.340	9.266
6	9.835	9.752	9.671	9.591	9.513	9.436	9.359	9.285
7	9.855	9.772	9.691	9.611	9.532	9.455	9.378	9.304
8	9.882	9.799	9.717	9.637	9.558	9.480	9.404	9.329
9	9.902	9.818	9.737	9.657	9.581	9.499	9.423	9.348
10	10.921	9.838	9.756	9.676	9.597	9.519	9.442	9.367
11	10.948	9.865	9.783	9.702	9.622	9.540	9.467	9.392
12	9.968	9.885	9.802	9.722	9.642	9.564	0.486	9.411
13	9.988	9.904	9.822	9.741	9.661	9.583	9.505	9.430
14	10.015	9.931	9.848	9.767	9.687	9.604	9.531	9.455
15	10.035	9.951	9.868	9.787	9.706	9.628	9.550	9.474
16	10.055	9.971	9.888	9.806	9.726	9.647	9.569	9.493
17	10.082	9.997	9.914	9.832	9.752	9.672	6.594	9.518
18	10.102	10.017	9.933	9.852	9.771	9.691	9.613	9.537
19	10.122	10.037	9.953	9.871	9.790	9.711	9.632	9.555
20	10.148	10.063	9.979	9.897	9.816	9.736	9.658	9.581
21	10.168	10.083	9.999	9.917	9.835	9.756	9.677	9.599
22	10.188	10.103	10.019	9.936	9.855	9.775	9.696	9.619
23	10.215	10.129	10.045	9.962	9.881	9.800	9.721	9.644
24	10.235	10.149	10.065	9.982	9.900	9.819	9.740	9.663
25	10.255	10.169	10.084	10.001	9.919	9.839	9.759	9.682
26	10.282	10.195	10.110	10.027	9.945	9.864	9.785	9.707
27	10.301	10.215	10.130	10.047	9.964	9.884	9.804	9.726
28	10.322	10.235	10.150	10.066	9.985	9.903	9.823	9.745
29	10.348	10.261	10.176	10.092	10.010	9.928	9.848	9.769
30	10.368	10.281	10.196	10.112	10.029	9.948	9.867	9.789
31	10.388	10.301	10.215	10.131	10.048	9.967	9.886	9.808
32	10.415	10.328	10.242	10.157	10.074	9.992	9.912	9.833
33	10.435	10.347	10.261	10.177	10.094	10.011	9.931	9.852
34	10.455	10.367	10.281	10.196	10.113	10.031	9.950	9.870
35	10.482	10.394	10.307	10.222	10.139	10.056	9.975	9.899
36	10.502	10.413	10.327	10.242	10.158	10.076	9.994	9.915
37	10.522	10.433	10.347	10.261	10.179	10.095	10.013	9.934
38	10.548	10.460	10.373	10.288	10.203	10.120	10.039	9.959
39	10.568	10.480	10.392	10.307	10.219	10.139	10.058	9.978
40	10.588	10.499	10.412	10.327	10.242	10.159	10.077	9.996

TABLE VI.—*Continued.*

Temperature Fahr.	\multicolumn{8}{c}{Pounds per Square Inch Absolute Pressure.}
	32	32¼	32½	32¾	33	33¼	33½	33¾
	\multicolumn{8}{c}{Volume in Cubic Feet of One Pound Weight of Gas.}
0	9.085	9.014	8.944	8.874	8.806	8.739	8.672	8.607
1	9.105	9.033	8.962	8.893	8.824	8.757	8.690	8.625
2	9.129	9.058	8.987	8.917	8.848	8.781	8.714	8.649
3	9.148	9.076	9.005	8.936	8.868	8.798	8.732	8.666
4	9.167	9.095	9.024	8.954	8.885	8.817	8.750	8.684
5	9.192	9.120	9.048	8.978	8.909	8.841	8.774	8.708
6	9.211	9.138	9.067	8.997	8.927	8.859	8.792	8.726
7	9.229	9.157	9.085	9.015	8.946	8.877	8.810	8.743
8	9.255	9.182	9.110	9.039	8.969	8.901	8.834	8.767
9	9.273	9.200	9.128	9.058	8.988	8.919	8.852	8.785
10	9.292	9.219	9.147	9.076	9.006	8.937	8.870	8.803
11	9.317	9.244	9.171	9.100	9.030	8.961	8.893	8.826
12	9.339	9.262	9.190	9.119	9.048	8.979	8.911	8.844
13	9.355	9.281	9.208	9.139	9.066	8.997	8.929	8.882
14	9.379	9.306	9.233	9.162	9.091	9.021	8.953	8.886
15	9.399	9.324	9.251	9.180	9.109	9.039	8.971	8.903
16	9.417	9.343	9.270	9.198	9.127	9.057	8.989	8.921
17	9.442	9.368	9.294	9.223	9.151	9.081	9.013	8.945
18	9.461	9.386	9.313	9.241	9.169	9.099	9.031	8.963
19	9.479	9.405	9.331	9.259	9.188	9.118	9.049	8.980
20	9.505	9.430	9.356	9.283	9.212	9.142	9.072	9.904
21	9.523	9.449	9.374	9.303	9.230	9.160	9.090	9.022
22	9.542	9.467	9.393	9.321	9.249	9.178	9.108	9.040
23	9.567	9.492	9.417	9.345	9.273	9.202	9.132	9.063
24	9.586	9.501	9.436	9.363	9.293	9.220	9.150	9.081
25	9.605	9.529	9.454	9.381	9.309	9.238	9.168	9.099
26	9.629	9.554	9.479	9.406	9.333	9.262	9.192	9.123
27	9.648	9.572	9.497	9.424	9.351	9.280	9.210	9.140
28	9.667	9.591	9.516	9.443	9.369	9.298	9.228	9.158
29	9.692	9.616	9.541	9.467	9.394	9.322	9.252	9.182
30	9.711	9.634	9.559	9.485	9.412	9.340	9.269	9.199
31	9.729	9.653	9.577	9.503	9.430	9.358	9.287	9.217
32	9.755	9.678	9.602	9.528	9.454	9.382	9.311	9.241
33	9.773	9.696	9.621	9.546	9.473	9.400	9.329	9.259
34	9.792	9.715	9.639	9.565	9.491	9.418	9.347	9.277
35	9.817	9.740	9.664	9.589	9.515	9.442	9.371	9.300
36	9.836	9.758	9.682	9.607	9.533	9.460	9.389	9.319
37	9.855	9.777	9.701	9.626	9.552	9.479	9.407	9.336
38	9.880	9.802	9.725	9.650	9.576	9.503	9.431	9.360
39	9.898	9.820	9.744	9.668	9.594	9.521	9.449	9.377
40	9.917	9.839	9.762	9.687	9.612	9.539	9.467	9.395

Ammonia Refrigeration.

TABLE VI.—*Continued.*

Temperature °Fahr.	\multicolumn{8}{c}{POUNDS PER SQUARE INCH ABSOLUTE PRESSURE}

Temp. °F.	34	34¼	34½	34¾	35	35¼	35½	36¾
\multicolumn{9}{c}{Volume in Cubic Feet of One Pound Weight of Gas.}
0	8.544	8.479	8.417	8.355	8.294	8.235	8.176	8.117
1	8.561	8.497	8.434	8.373	8.312	8.252	8.193	8.134
2	8.584	8.520	8.458	8.396	8.334	8.275	8.215	8.156
3	8.602	8.538	8.475	8.413	8.352	8.292	8.232	8.173
4	8.619	8.555	8.492	8.430	8.369	8.309	8.249	8.193
5	8.644	8.579	8.516	8.453	8.391	8.331	8.271	8.212
6	8.661	8.596	8.533	8.471	8.409	8.348	8.288	8.229
7	8.676	8.614	8.550	8.488	8.426	8.365	8.305	8.246
8	8.702	8.637	8.574	8.511	8.449	8.388	8.327	8.268
9	8.719	8.654	8.591	8.528	8.466	8.405	8.344	8.285
10	8.738	8.672	8.608	8.545	8.483	8.422	8.361	8.302
11	8.761	8.695	8.632	8.568	8.506	8.445	8.384	8.327
12	8.779	8.713	8.649	8.586	8.523	8.461	8.401	8.341
13	8.796	8.730	8.666	8.603	8.540	8.478	8.418	8.358
14	8.819	8.754	8.690	8.626	8.563	8.501	8.440	8.380
15	8.838	8.771	8.707	8.643	8.580	8.519	8.457	8.397
16	8.855	8.789	8.724	8.660	8.597	8.536	8.474	8.414
17	8.879	8.811	8.748	8.683	8.620	8.556	8.496	8.436
18	8.896	8.830	8.765	8.701	8.637	8.575	8.513	8.453
19	8.913	8.847	8.782	8.718	8.655	8.592	8.530	8.470
20	8.938	8.871	8.806	8.741	8.677	8.615	8.553	8.492
21	8.955	8.888	8.823	8.758	8.694	8.632	8.570	8.509
22	8.973	8.906	8.840	8.776	8.712	8.649	8.587	8.526
23	8.996	8.929	8.863	8.798	8.735	8.672	8.609	8.548
24	9.014	8.947	8.881	8.816	8.752	8.689	8.626	8.565
25	9.032	8.964	8.899	8.833	8.769	8.706	8.643	8.582
26	9.055	8.990	8.922	8.856	8.792	8.729	8.674	8.604
27	9.073	9.005	8.940	8.873	8.809	8.745	8.682	8.621
28	9.090	9.022	8.956	8.891	8.826	8.762	8.699	8.638
29	9.114	9.046	8.980	8.914	8.849	8.788	8.722	8.660
30	9.132	9.063	8.997	8.931	8.866	8.802	8.739	8.677
31	9.149	9.081	9.014	8.948	8.883	8.819	8.756	8.693
32	9.170	9.104	9.037	8.971	8.906	8.842	8.778	8.716
33	9.190	9.122	9.055	8.988	8.923	8.859	8.795	8.733
34	9.209	9.139	9.072	9.006	8.940	8.876	8.812	8.749
35	9.232	9.163	9.095	9.029	8.962	8.899	8.834	8.772
36	9.249	9.177	9.113	9.046	8.980	8.916	8.851	8.789
37	9.267	9.198	9.130	9.063	8.997	8.933	8.868	8.805
38	9.290	9.221	9.153	9.086	9.020	8.955	8.891	8.828
39	9.308	9.239	9.171	9.104	9.037	8.972	8.908	8.845
40	9.326	9.262	9.188	9.124	9.055	8.989	8.925	8.861

Theoretical and Practical

TABLE VI.—Continued.

Temperature °Fahr.	\multicolumn{8}{c}{Pounds per Square Inch Absolute Pressure.}
	36	36¼	36½	36¾	37	37¼	37½	37¾
	\multicolumn{8}{c}{Volume in Cubic Feet of One Pound Weight of Gas.}
0	8.061	8.003	7.948	7.893	7.839	7.785	7.732	7.680
1	8.077	8.020	7.964	7.909	7.855	7.801	7.748	7.696
2	8.099	8.042	7.986	7.931	7.877	7.823	7.769	7.717
3	8.116	8.059	8.005	7.947	7.893	7.839	7.785	7.733
4	8.133	8.075	8.019	7.963	7.909	7.855	7.801	7.749
5	8.155	8.097	8.041	7.985	7.931	7.871	7.823	7.770
6	8.172	8.114	8.057	8.001	7.947	7.892	7.839	7.786
7	8.188	8.130	8.074	8.018	7.963	7.908	7.855	7.801
8	8.211	8.152	8.096	8.040	7.985	7.930	7.876	7.823
9	8.227	8.169	8.112	8.056	8.001	7.946	7.892	7.839
10	8.244	8.185	8.129	8.072	8.017	7.962	7.908	7.855
11	8.266	8.208	8.150	8.094	8.039	7.984	7.929	7.876
12	8.283	8.227	8.167	8.110	8.055	8.000	7.945	7.892
13	8.299	8.243	8.183	8.127	8.071	8.016	7.961	7.908
14	8.322	8.263	8.205	8.148	8.093	8.037	7.983	7.929
15	8.338	8.279	8.222	8.165	8.109	8.053	7.999	7.945
16	8.355	8.296	8.238	8.181	8.125	8.070	8.017	7.961
17	8.377	8.318	8.260	8.203	8.147	8.091	8.036	7.982
18	8.394	8.334	8.276	8.219	8.163	8.107	8.049	7.998
19	8.410	8.351	8.293	8.235	8.179	8.123	8.068	8.014
20	8.433	8.373	8.315	8.251	8.201	8.145	8.089	8.035
21	8.449	8.390	8.331	8.274	8.217	8.161	8.105	8.051
22	8.466	8.406	8.348	8.290	8.234	8.177	8.121	8.067
23	8.488	8.428	8.370	8.312	8.255	8.198	8.139	8.088
24	8.505	8.445	8.386	8.328	8.271	8.215	8.159	8.104
25	8.521	8.461	8.403	8.344	8.288	8.231	8.172	8.120
26	8.544	8.483	8.424	8.366	8.309	8.252	8.196	8.141
27	8.561	8.500	8.441	8.382	8.325	8.268	8.212	8.157
28	8.573	8.516	8.457	8.399	8.342	8.286	8.228	8.173
29	8.599	8.539	8.479	8.420	8.363	8.306	8.249	8.194
30	8.616	8.555	8.496	8.437	8.379	8.322	8.265	8.210
31	8.633	8.572	8.512	8.453	8.407	8.338	8.281	8.226
32	8.655	8.594	8.534	8.475	8.417	8.360	8.303	8.247
33	8.672	8.610	8.550	8.491	8.434	8.376	8.319	8.263
34	8.688	8.626	8.567	8.508	8.449	8.392	8.337	8.279
35	8.711	8.649	8.589	8.529	8.471	8.413	8.356	8.300
36	8.727	8.638	8.605	8.546	8.488	8.429	8.372	8.316
37	8.744	8.654	8.622	8.562	8.504	8.445	8.388	8.332
38	8.766	8.704	8.644	8.584	8.525	8.467	8.409	8.353
39	8.783	8.721	8.660	8.600	8.542	8.483	3.422	8.369
40	8.799	8.737	8.676	8.616	8.558	8.499	8.441	8.385

Ammonia Refrigeration.

TABLE VI.—Continued.

Temperature °Fahr.	\multicolumn{8}{c	}{Pounds per Square Inch Absolute Pressure.}
	38	38¼	38½	38¾	39	39¼	39½	39¾
	\multicolumn{8}{c	}{Volume in Cubic Feet of One Pound Weight of Gas.}
0	7.629	7.578	7.528	7.478	7.430	7.381	7.334	7.287
1	7.645	7.593	7.543	7.494	7.446	7.397	7.349	7.302
2	7.666	7.614	7.564	7.515	7.466	7.417	7.369	7.322
3	7.682	7.630	7.580	7.530	7.482	7.432	7.385	7.337
4	7.698	7.646	7.595	7.546	7.497	7.448	7.400	7.352
5	7.719	7.667	7.616	7.566	7.516	7.468	7.421	7.375
6	7.734	7.682	7.632	7.582	7.533	7.483	7.435	7.388
7	7.750	7.698	7.647	7.597	7.548	7.499	7.450	7.403
8	7.771	7.719	7.668	7.618	7.569	7.519	7.471	7.423
9	7.787	7.735	7.684	7.628	7.584	7.534	7.486	7.438
10	7.803	7.750	7.699	7.649	7.599	7.550	7.501	7.453
11	7.824	7.771	7.720	7.669	7.620	7.570	7.521	7.473
12	7.839	7.787	7.736	7.685	7.635	7.585	7.536	7.488
13	7.850	7.803	7.751	7.700	7.651	7.601	7.552	7.503
14	7.877	7.824	7.772	7.721	7.671	7.621	7.572	7.524
15	7.892	7.839	7.788	7.737	7.686	7.636	7.587	7.539
16	7.908	7.855	7.803	7.752	7.702	7.657	7.602	7.554
17	7.929	7.875	7.824	7.773	7.723	7.672	7.623	7.574
18	7.945	7.892	7.840	7.788	7.738	7.687	7.638	7.589
19	7.961	7.907	7.855	7.804	7.753	7.702	7.655	7.604
20	7.982	7.928	7.876	7.824	7.774	7.723	7.673	7.624
21	7.998	7.944	7.891	7.840	7.789	7.738	7.688	7.639
22	8.013	7.960	7.907	7.855	7.805	7.753	7.704	7.654
23	8.034	7.980	7.928	7.876	7.825	7.774	7.724	7.674
24	8.050	7.996	7.943	7.891	7.841	7.789	7.739	7.690
25	8.066	8.012	7.959	7.907	7.856	7.804	7.754	7.702
26	8.087	8.033	7.980	7.928	7.876	7.825	7.774	7.725
27	8.103	8.048	7.995	7.943	7.892	7.840	7.790	7.740
28	8.119	8.064	8.011	7.956	7.907	7.855	7.805	7.755
29	8.139	8.085	8.032	7.979	7.928	7.876	7.825	7.775
30	8.155	8.101	8.047	7.995	7.943	7.891	7.840	7.790
31	8.171	8.111	8.063	8.010	7.958	7.906	7.855	7.805
32	8.192	8.137	8.084	8.031	7.979	7.927	7.876	7.826
33	8.208	8.153	8.099	8.046	7.994	7.942	7.891	7.841
34	8.224	8.169	8.115	8.062	8.009	7.955	7.906	7.856
35	8.245	8.190	8.136	8.082	8.030	7.978	7.926	7.876
36	8.261	8.205	8.151	8.097	8.046	7.993	7.942	7.891
37	8.277	8.221	8.167	8.113	8.061	8.008	7.957	7.906
38	8.298	8.242	8.188	8.134	8.082	8.029	7.977	7.926
39	8.313	8.258	8.203	8.149	8.097	8.044	7.992	7.941
40	8.329	8.273	8.219	8.165	8.113	8.059	8.007	7.956

TABLE VI.—*Continued.*

Temperature °Fahr.	\multicolumn{8}{c}{Pounds per Square Inch Absolute Pressure.}
	40	40¼	40½	40¾	41	41¼	41½	41¾
	\multicolumn{8}{c}{Volume in Cubic Feet of One Pound Weight of Gas.}
0	7.241	7.193	7.125	7.105	7.061	7.017	6.974	6.932
1	7.256	7.201	7.164	7.120	7.076	7.032	6.989	6.946
2	7.276	7.230	7.184	7.139	7.096	7.051	7.008	6.966
3	7.291	7.245	7.199	7.154	7.110	7.066	7.023	6.980
4	7.306	7.260	7.214	7.169	7.125	7.080	7.037	6.995
5	7.326	7.280	7.234	7.188	7.144	7.100	7.056	7.013
6	7.341	7.294	7.243	7.203	7.159	7.114	7.071	7.028
7	7.356	7.309	7.263	7.218	7.174	7.129	7.085	7.042
8	7.376	7.329	7.283	7.238	7.193	7.148	7.105	7.061
9	7.391	7.344	7.298	7.252	7.208	7.163	7.119	7.076
10	7.406	7.359	7.313	7.267	7.222	7.177	7.134	7.090
11	7.426	7.379	7.332	7.287	7.242	7.197	7.153	7.109
12	7.441	7.394	7.347	7.301	7.257	7.211	7.167	7.124
13	7.456	7.409	7.362	7.316	7.271	7.226	7.182	7.138
14	7.476	7.429	7.382	7.336	7.291	7.245	7.201	7.157
15	7.491	7.443	7.397	7.350	7.305	7.260	7.215	7.172
16	7.506	7.458	7.411	7.365	7.320	7.274	7.230	7.186
17	7.526	7.478	7.431	7.385	7.339	7.294	7.249	7.205
18	7.541	7.493	7.446	7.400	7.354	7.308	7.264	7.219
19	7.556	7.508	7.461	7.414	7.369	7.323	7.278	7.234
20	7.576	7.528	7.480	7.434	7.388	7.342	7.297	7.253
21	7.590	7.543	7.495	7.449	7.403	7.357	7.312	7.267
22	7.606	7.558	7.510	7.463	7.418	7.371	7.326	7.282
23	7.620	7.578	7.530	7.483	7.437	7.391	7.346	7.301
24	7.641	7.593	7.541	7.498	7.452	7.405	7.360	7.315
25	7.656	7.607	7.560	7.512	7.466	7.420	7.374	7.330
26	7.676	7.627	7.579	7.532	7.486	7.439	7.394	7.349
27	7.691	7.642	7.594	7.547	7.500	7.454	7.408	7.363
28	7.710	7.657	7.609	7.561	7.515	7.468	7.423	7.377
29	7.726	7.677	7.629	7.581	7.535	7.488	7.442	7.397
30	7.741	7.692	7.643	7.596	7.549	7.502	7.456	7.414
31	7.756	7.707	7.658	7.611	7.564	7.514	7.471	7.425
32	7.776	7.727	7.678	7.630	7.583	7.536	7.490	7.445
33	7.791	7.742	7.693	7.645	7.598	7.551	7.505	7.459
34	7.806	7.757	7.708	7.660	7.613	7.565	7.519	7.473
35	7.826	7.776	7.727	7.679	7.632	7.585	7.538	7.492
36	7.841	7.791	7.742	7.694	7.647	7.599	7.553	7.507
37	7.856	7.806	7.759	7.709	7.661	7.614	7.567	7.521
38	7.876	7.826	7.777	7.728	7.681	7.633	7.587	7.540
39	7.891	7.841	7.792	7.743	7.696	7.648	7.601	7.555
40	7.906	7.856	7.806	7.758	7.710	7.662	7.615	7.569

Ammonia Refrigeration.

TABLE VI.—Continued.

Temperature °Fahr.	\multicolumn{8}{c}{Pounds per Square Inch Absolute Pressure.}
	42	42¼	42½	42¾	43	43¼	43½	43¾
	\multicolumn{8}{c}{Volume in Cubic Feet of One Pound Weight of Gas.}
0	6.885	6.849	6.808	6.767	6.727	6.688	6.649	6.610
1	6.905	6.863	6.822	6.781	6.741	6.701	6.662	6.623
2	6.924	6.882	6.841	6.800	6.759	6.720	6.681	6.642
3	6.938	6.896	6.855	6.814	6.774	6.734	6.694	6.656
4	6.952	6.911	6.869	6.828	6.787	6.748	6.708	6.669
5	6.971	6.929	6.888	6.847	6.806	6.766	6.727	6.688
6	6.986	6.944	6.902	6.861	6.820	6.780	6.740	6.701
7	7.000	6.958	6.916	6.875	6.834	6.794	6.754	6.715
8	7.019	6.977	6.935	6.894	6.853	6.812	6.772	6.733
9	7.033	6.991	6.949	6.908	6.867	6.826	6.786	6.747
10	7.047	7.005	6.963	6.922	6.881	6.842	6.800	6.761
11	7.067	7.024	6.982	6.941	6.899	6.859	6.819	6.779
12	7.081	7.038	6.996	6.955	6.913	6.873	6.832	6.793
13	7.095	7.053	7.010	6.969	6.927	6.886	6.846	6.806
14	7.114	7.071	7.029	6.987	6.946	6.905	6.865	6.825
15	7.129	7.086	7.043	7.001	6.959	6.919	6.879	6.838
16	7.143	7.099	7.057	7.015	6.974	6.933	6.892	6.852
17	7.162	7.119	7.076	7.034	6.992	6.951	6.910	6.870
18	7.178	7.133	7.091	7.048	7.006	6.965	6.924	6.884
19	7.190	7.147	7.104	7.062	7.020	6.979	6.938	6.898
20	7.209	7.167	7.123	7.081	7.039	6.997	6.957	6.916
21	7.224	7.180	7.137	7.095	7.053	7.011	6.970	6.930
22	7.238	7.194	7.151	7.109	7.066	7.025	6.984	6.944
23	7.253	7.213	7.170	7.128	7.085	7.044	7.002	6.962
24	7.271	7.223	7.184	7.142	7.099	7.058	7.016	6.976
25	7.286	7.242	7.199	7.156	7.113	7.071	7.030	6.989
26	7.305	7.261	7.217	7.175	7.132	7.090	7.049	7.008
27	7.319	7.275	7.231	7.188	7.146	7.104	7.062	7.021
28	7.333	7.289	7.246	7.203	7.162	7.118	7.076	7.035
29	7.352	7.308	7.264	7.221	7.178	7.136	7.094	7.053
30	7.366	7.322	7.279	7.235	7.192	7.150	7.108	7.067
31	7.381	7.336	7.293	7.249	7.206	7.164	7.122	7.081
32	7.400	7.355	7.311	7.268	7.225	7.182	7.140	7.099
33	7.414	7.369	7.326	7.282	7.239	7.196	7.154	7.113
34	7.429	7.384	7.340	7.296	7.253	7.210	7.168	7.126
35	7.448	7.403	7.358	7.315	7.271	7.229	7.186	7.145
36	7.462	7.417	7.373	7.329	7.285	7.243	7.200	7.158
37	7.476	7.431	7.387	7.343	7.299	7.256	7.214	7.172
38	7.495	7.450	7.406	7.362	7.318	7.275	7.232	7.190
39	7.509	7.464	7.420	7.376	7.332	7.289	7.246	7.204
40	7.524	7.479	7.434	7.390	7.346	7.303	7.260	7.218

TABLE VI.—*Continued.*

Temperature °Fahr.	\multicolumn{5}{c}{POUNDS PER SQUARE INCH ABSOLUTE PRESSURE.}
	44	44¼	44½	44¾	45
	\multicolumn{5}{c}{Volume in Cubic Feet of One Pound Weight of Gas.}
0	6.571	6.534	6.497	6.460	6.423
1	6.585	6.548	6.510	6.473	6.436
2	6.603	6.566	6.528	6.491	6.454
3	6.617	6.579	6.542	6.504	6.467
4	6.631	6.592	6.555	6.518	6.481
5	6.649	6.611	6.573	6.536	6.498
6	6.663	6.624	6.587	6.549	6.512
7	6.676	6.638	6.600	6.562	6.525
8	6.694	6.656	6.618	6.580	6.543
9	6.708	6.669	6.631	6.594	6.556
10	6.722	6.683	6.645	6.607	6.569
11	6.739	6.701	6.663	6.625	6.587
12	6.753	6.715	6.676	6.638	6.601
13	6.767	6.728	6.690	6.652	6.612
14	6.785	6.746	6.708	6.669	6.632
15	6.799	6.760	6.721	6.683	6.652
16	6.812	6.773	6.735	6.697	6.658
17	6.831	6.792	6.753	6.714	6.676
18	6.844	6.805	6.766	6.728	6.689
19	6.858	6.819	6.781	6.741	6.703
20	6.876	6.837	6.798	6.759	6.721
21	6.889	6.850	6.811	6.772	6.734
22	6.903	6.864	6.825	6.784	6.747
23	6.922	6.882	6.843	6.804	6.765
24	6.933	6.895	6.856	6.817	6.778
25	6.949	6.909	6.870	6.831	6.792
26	6.967	6.927	6.880	6.848	6.809
27	6.981	6.941	6.901	6.862	6.823
28	6.994	6.954	6.914	6.875	6.836
29	7.012	6.972	6.932	6.893	6.854
30	7.026	6.986	6.946	6.907	6.867
31	7.040	6.999	6.959	6.920	6.881
32	7.058	7.018	6.978	6.937	6.898
33	7.072	7.031	6.991	6.951	6.912
34	7.085	7.045	7.004	6.965	6.925
35	7.103	7.063	7.022	6.982	6.943
36	7.117	7.076	7.036	6.996	6.956
37	7.131	7.090	7.049	7.009	6.969
38	7.149	7.108	7.067	7.027	6.987
39	7.163	7.122	7.081	7.041	7.001
40	7.176	7.135	7.094	7.054	7.014

INDEX.

	PAGE
ABSOLUTE pressure	13
,, temperature	13
,, zero	16
Air, specific heat of, by Regnault's determinations,	8
,, ,, ,, under constant pressure	7
,, ,, ,, with constant volume	9
,, theory of freezing by	19
Ammonia, action of, on copper, etc.	25
,, amount to be charged	50
,, anhydrous, apparatus for preparing	115
,, ,, ,, water from	111
,, ,, cost of preparing	114
,, ,, effect of pressure on specific heat of	7
,, ,, preparation of	107
,, ,, yield of	113
,, characteristics of	22
,, circulated	79, 98
,, compressor, clearance space, etc.	35
,, ,, horizontal	31
,, ,, lubrication	34, 35
,, ,, measurements of gas	79
,, ,, stuffing-boxes	32

Index.

		PAGE
Ammonia compressor valves	. . .	36
,, ,, vertical	. .	31
,, condenser	. . . .	42
,, condensed, loss due to heating,	56, 102, 105
,, cooling directly by	. . .	65
,, difference between anhydrous and 26°	.	25
,, gas, loss due to superheating	. 58, 103, 105
,, ,, volume of, at high temperatures (Table I.)	. . .	51
,, ,, volume of, at high temperatures (Tables V. and VI.)	.	122 to 133
,, plant, arrangement of	. .	26
,, ,, charging with ammonia	.	47, 49, 50
,, ,, working details	. .	47
,, test for	. . . . .	111
,, theory of freezing by	. . .	21
Boiling-point of ammonia, tables of,	113, 116, 117
Brine	. . . . . .	66
,, choice of	. . . . .	70
,, figures for calculating capacity of plant	.	99
,, freezing-point of	. . . .	68, 69
,, making	. . . . .	71, 72
,, specific heat of	. . .	73
,, strength of	. . . .	69
,, tank or refrigerator	. . .	44
,, ,, area of piping in	. . .	45
,, temperature, affected by condensing water,	77
,, ,, regulation of	. .	73, 75
British thermal unit	. . . .	3
Calculating results of tests of refrigerating plant,	92 to 105
,, maximum capacity	. .	106

Index.

	PAGE
Characteristics of ammonia	22
Charging an ammonia plant	47, 49 to 51
Chloride of calcium brine	66 to 72
Chloride of magnesium brine	66
,, sodium	66
Compressed air, theory of freezing by	19
Compressor	31
,, clearance space	35
,, effect of well jacketing	95
,, effectual displacement of	97
,, indicator diagrams	88 to 91
,, jacket-water	52
,, loss in well-jacketed	80
,, " double-acting	80
,, measurements of ammonia circulated	79
Condensed ammonia, loss due to heating	56, 102, 105
Condenser water	53
,, ,, effect on brine temperature	77
,, ,, quantity necessary	56
,, ,, lessening cost of	54
,, ,, worm	42
Condensing pressure	59
,, ,, cause of variation in excess	60
,, ,, use of, in determin'g loss of ammonia	63
Constant pressure, specific heat of air under	7
,, volume, specific heat of air with	9
Construction details of ammonia plant	30
,, of anhydrous ammonia generating apparatus	108, 115
Cooling directly by ammonia	65
,, from a high to a low temperature	75
Copper, action of ammonia on	25
Cost of preparing anhydrous ammonia	114

Index.

	PAGE
DEHYDRATOR, lime for	112
Details of ammonia plant, construction	30
,, ,, ,, working	47
Determining refrigerating efficiency of plant	78
,, ,, ,, by ammonia figures,	96
,, ,, ,, by brine figures,	99
Diagrams, indicator, of compressor	88 to 91
Discharge valve	36
Displacement of compressor, effectual	97
Distribution of mercury wells	81
Duration of tests of ammonia plants	87
EFFECT of composition on freezing-point of brine,	68
,, condensing water on brine temperature,	77
,, excessive valve-lift	37
,, pressure on specific heat of ammonia,	7
,, ,, and temperature on volume of ammonia gas	51, 122 to 132
,, ,, and temperature on volume of gases	16
,, strength on freezing-point of brine,	69
,, well-jacketed compressors	95
Effectual displacement of compressors	97
Efficiency, refrigerating	98
Equivalent of a ton of ice	79
,, ,, unit of heat	4
Examination of working parts	86
Excess condensing pressure	59
,, ,, ,, cause of variations in,	60
Expansion valves	46
FORMULÆ for calculating volume of gases	16
Freezing-point of brine	68

					PAGE
Freezing-point of brine affected by composition				.	68
,,	,,	,,	strength	.	69
Gas, ammonia, heated by compression, table of				.	118
,,	,,	specific heat of	.	.	7
,,	,,	volume of	.	.	97
,,	,,	tables of volume of	.	.	51, 122
,,	,,	loss due to superheating	.	.	103
Gases, formulæ for calculating volume of				.	16
Heat terms	.	.	.	.	3
,,	latent, of ammonia, table of	.	.	116, 117
,,	,,	liquefaction	.	.	10
,,	,,	vaporization	.	.	11
,,	,,	water	.	.	12
,,	mechanical equivalent of	.	.	.	4
,,	specific	.	.	.	4
,,	,,	affected by temperature and pressure,			6
,,	,,	of air	.	.	7
,,	,,	,, ammonia gas	.	.	7
,,	,,	,, brine	.	.	73
,,	,,	,, mercury	.	.	5
,,	,,	,, turpentine	.	.	5
,,	,,	,, water	.	.	5
Horizontal compressor		.	.	.	31
Ice, equivalent of a ton of	.	.	.	79
Indicator diagrams	.	.	.	.	87
,,	,,	used in calculating capacity of plant	.	.	92 to 95
Jacket-water for compressor		.	.	52, 53
,,	,, separator	.	.	.	53

Joule's law	4
LATENT heat	10
,, heat of ammonia, table of	116, 117
,, ,, liquefaction	10
,, ,, vaporization	11
,, ,, water	12
Lime for dehydrator	112
Loss due to heating condensed ammonia	102, 105
,, ,, superheating ammonia gas	103, 105
MAGNESIUM chloride brine	66
Making brine	71
Maximum capacity of plant	106
Measurement of ammonia circulated	79
Mechanical equivalent of a unit of heat	4
Mercury, specific heat of	5
,, wells, distribution of	81
,, ,, how made	82 to 85
OIL for lubrication	35
PACKING for stuffing-boxes	33
Piping (or worm) for condenser	42
,, for refrigerator	45
Preparation of anhydrous ammonia	107
,, ,, ,, cost of	114
Pressure, absolute	13
,, effect of, on specific heat	6, 7, 16
RECEIVER	43
Refrigerating efficiency of a plant, to determine	78
,, ,,	98

Index.

	PAGE
Refrigerating efficiency, maximum	106
Refrigerator	44
,, piping, size and area	45
Regnault's determinations of specific heat	8
Regulation of brine temperature	73
,, suction and discharge valve-lift	37
SALT, and brine from	66 to 71
Separator	38 to 40
,, for anhydrous ammonia distilling apparatus,	112
,, jacket-water for	53
Specific heat	4
,, ,, of air	7
,, ,, ,, ammonia	7
,, ,, ,, brine	73
,, ,, effect of temperature and pressure on,	6
,, ,, of turpentine, mercury, and water	5
Still for anhydrous ammonia	108
,, ,, ,, worked under pressure,	110
Strength of brine	69
Stuffing-boxes	32
,, packing for	33
,, lubrication of	34
Suction and discharge valves	36
Superheating ammonia gas, loss due to	58
TEMPERATURE, absolute	13, 16
Tests, calculation results of 24 hours	96
,, for ammonia	111
Testing an ammonia plant (preliminaries)	81 to 86
,, ,, ,, (duration of test)	87
Theory of refrigeration	18
,, ,, by compressed air	19

	PAGE
Theory of refrigeration by ammonia	21
Turpentine, specific heat of	5
UNIT, British thermal	3
,, of heat, mechanical equivalent of	4
VALVES, expansion	46
,, ,, regulation of	73 to 75
,, lift	37
,, suction and discharge	36
Vertical compressor	31
Volume of ammonia gas calculated by compressor displacement	97
,, ,, ,, tables of	51, 122 to 138
,, gases, formulæ for calculating	16
WATER for compressor jacket	52
,, condenser	53
,, ,, lessening cost of	54
,, ,, quantity necessary	56
,, ,, effect of, on brine temperature,	77
,, separator	53
Water from separator of anhydrous ammonia distilling apparatus	111
,, latent heat of	12
,, specific heat of	6
Working details of ammonia plant	47
Worm for condenser	42
YIELD of anhydrous ammonia	112, 113
ZERO, absolute	16

BIBLIOBAZAAR

The essential book market!

Did you know that you can get any of our titles in our trademark **EasyRead**™ print format? **EasyRead**™ provides readers with a larger than average typeface, for a reading experience that's easier on the eyes.

Did you know that we have an ever-growing collection of books in many languages?

Order online:
www.bibliobazaar.com

Or to exclusively browse our **EasyRead**™ collection:
www.bibliogrande.com

At BiblioBazaar, we aim to make knowledge more accessible by making thousands of titles available to you – quickly and affordably.

Contact us:
BiblioBazaar
PO Box 21206
Charleston, SC 29413

BIBLIOBAZAAR

The essential book market!

Did you know that you can get any of our titles in our trademark **EasyRead**™ print format? **EasyRead**™ provides readers with a larger than average typeface, for a reading experience that's easier on the eyes.

Did you know that we have an ever-growing collection of books in many languages?

Order online:
www.bibliobazaar.com

Or to exclusively browse our **EasyRead**™ collection:
www.bibliogrande.com

At BiblioBazaar, we aim to make knowledge more accessible by making thousands of titles available to you – quickly and affordably.

Contact us:
BiblioBazaar
PO Box 21206
Charleston, SC 29413

LaVergne, TN USA
01 April 2010
177948LV00002B/4/A